Robert Paul and the
Origins of British Cinema

Cinema and Modernity

Edited by Tom Gunning

Robert Paul and the Origins of British Cinema

Ian Christie

The University of Chicago Press CHICAGO AND LONDON

The University of Chicago Press, Chicago 60637
The University of Chicago Press, Ltd., London
© 2019 by Ian Christie
Published 2019
Printed in the United States of America

28 27 26 25 24 23 22 21 20 19   1 2 3 4 5

ISBN-13: 978-0-226-10562-8 (cloth)
ISBN-13: 978-0-226-10563-5 (paper)
ISBN-13: 978-0-226-61011-5 (e-book)
DOI: https://doi.org/10.7208/chicago/9780226610115.001.0001

Library of Congress Cataloging-in-Publication Data

Names: Christie, Ian, 1945– author.
Title: Robert Paul and the origins of British cinema /
Ian Christie.
Other titles: Cinema and modernity.
Description: Chicago : University of Chicago Press, 2019. |
Series: Cinema and modernity | Includes bibliographical
references and index.
Identifiers: LCCN 2019032253 | ISBN 9780226105628 (cloth) |
ISBN 9780226105635 (paperback) | ISBN 9780226610115 (ebook)
Subjects: LCSH: Paul, Robert, 1869-1943. | Cinematographers —
England — Biography. | Electrical engineers — England —
Biography. | Cinematography — England — History. |
Motion picture projectors — England — History. | Motion
picture industry — England — History.
Classification: LCC TR849.P38 C47 2019 |
DDC 777/.8092 [B] — dc23
LC record available at https://lccn.loc.gov/2019032253

♾ This paper meets the requirements of ANSI/NISO Z39.48-1992
(Permanence of Paper).

This book is dedicated to the memory of John Barnes (1920–2008), the pioneer historian of early British cinema, who inspired and encouraged me, although sadly did not live to see it completed

Contents

Acknowledgments

Any book that has been under way for so long will accumulate more debts that can properly be acknowledged. The starting point for this one was *The Last Machine*, a book and television series commissioned by BBC Television for the centenary of cinema, which appeared in 1994. In this, after Colin MacCabe opened the door, John Wyver and Richard Curson Smith first guided me toward making early film history accessible. Later, David Thompson included me discussing Robert Paul in his BBC film *Silent Britain*; Tom Gunning welcomed the book for his series at the University of Chicago Press; and a senior fellowship from the Paul Mellon Centre for British Art released me from teaching at Birkbeck and helped research get started. At the British Film Institute, Heather Stewart ensured that a DVD was produced of Paul's surviving films, several of which were reconstituted from Filoscope copies held by what is now the Bill Douglas Cinema Museum at Exeter University and what has since become the National Science and Media Museum in Bradford. Other material came from what are now the George Eastman Museum (Paolo Cherchi Usai, Caroline Yeager) and Ngi Taonga New Zealand Archive of Film, Television and Sound (Bronwyn Taylor, Cushla Vula), as well as the Scottish Screen Archive (Janet McBain) and the National Fairground and Circus Archive, University of Sheffield (Vanessa Toulmin). Archivists Camille Blot-Wellens at the Swedish Film Institute and Bryony Dixon at the BFI National Archive have continued to add to the stock of known Paul films during the time of my writing.

Over the years, I have enjoyed generous support from many within the early film archival and research community and beyond. The biannual Domitor Conferences and British Silents conferences and symposia have heard portions of my research as papers, some published in their proceedings. Paul Hammond readily gave me access to his research on Walter Booth; Stephen Bottomore and Simon Popple both provided me with the fruits of their findings in early journals; and Stephen Herbert has been a frequent source of wise advice, as well as first alerting me to Irene Codd as a potentially valuable source. In Stockholm, Jan Olsson patiently helped me investigate Paul's screenings and contacts with the royal family. And from Norway Jan Anders Diesen kindly sent me the photograph of Paul taken by Ottar Gladtvet, from his 1999 book on this Norwegian pioneer, *Filmeventyret beginner* (The film adventure begins). I am grateful to Laila Johns of the Norwegian National Library for confirming details of this. Patrick Parrinder, doyen of Wells scholars, has kindly given up-to-date advice on continuing scholarship about Wells's writings in 1898–1900. Laurent Mannoni, scholar of early cinema, and curator and conservator at the Cinématheque Française, gave me access to the Will Day Collection. Jean-Jacques Meusy has been an invaluable advisor on aspects of early French cinema. Roland-François Lack has generously shared his own research on Paul in relation to our common location in North London. John and Maggie Dodd, current owners of Paul's house on Sydney Road, generously invited me in to see the original features they have preserved. Also in North London, I was encouraged by the late Ken Gay, local historian of Muswell Hill, and benefited from consulting the paper collection of the Hornsey Historical Society. On Paul's instrument-making business, Erasmus Barlow, onetime chairman of the Cambridge Scientific Instrument Company, granted me an interview about the firm's history, and Don Unwin, a former employee, recalled the ethos of the company, showing me Paul's own milling machine. But my largest debt is to Richard Brown, intrepid independent scholar, who has aided my research on Paul over many years, generously sharing his own discoveries, despite his regrettably low opinion of Paul's motives and veracity. Without his diligence and skills, despite our disagreements, this would be a poorer book.

Members of my family have had to live with Paul for what must seem much of their lives. My sister-in-law and her husband, Christine and Kim Murchison, kindly housed me in Sydney during a research visit. My daughter Laura Christie was especially helpful in searching for evidence of Paul's family. And above all, my wife, Patsy Nightingale, has never lost faith that there would be an outcome we could be proud of, for which I am more in her debt than ever.

1. Early publicity portrait, with Paul's signature (1890s).

Prelude

Why should the beginning of moving pictures in the mid-1890s matter, except to a small coterie of specialist film and media historians? A common answer is that from such short and often jerky beginnings as the Kinetoscope parlor and the first Cinématographe shows would emerge a mighty empire of fantastic fiction that has dominated the world's imagination for the last century. And that the underpinning of this empire, in a series of interlocking global businesses, continues to shape the political economy of the world we live in. Fox, Columbia, Warner, Paramount, Universal—these names that link the mythic world of the early Hollywood studios with the multimedia and multinational entertainment businesses of today—are, as the story goes, direct heirs of the pioneers who first launched moving pictures: Edison, the Lumières, and Robert Paul.

Essentially the same story could be told of any of the industries that forged the modern world, whether we consider the pervasive impact of electric power, telecommunications, and chemical engineering, or new means of mobility such as bicycles, automobiles, and aircraft, or devices such as the sewing machine, the phonograph, and the typewriter. From experimental beginnings, between roughly 1880 and 1910, these would have an accelerating impact on the lives and livelihood of citizens of the advanced countries; indeed, the extent of their impact would quickly come to define what counted as "advanced." That assessment rests upon a raft of assumptions about technology, culture, and "consciousness"—assumptions that have been persuasively tied together by such varied theorists as Harold Innis, his disciple Marshall McLuhan, Theodor Adorno, and Friedrich Kittler, but also challenged by those who resist the seductive idea of a decisive fin-de-siècle watershed.

Moving pictures, which would soon grow into cinema, play an ambiguous role in these debates. Were they a symptom or a cause of the complex transformation we label "modernity"? A major or a minor player?

These are not questions that concerned most cinema historians during the twentieth century, who told a confident story about the emergence of a new medium and its rise to maturity as the equal of such old media as the novel and dramatic theater. But such histories have in-built limitations, however eloquent and extensive they may be. They tend to isolate cinema from its immediate hinterland and to stress its distinctness, so magnifying its importance. Now, however, with a growing awareness of cinema as one of a range of moving-image entertainment media, has come a growing interest in its origins—within a matrix of "new media" at the turn of the twentieth century that was eerily similar to our present situation.

What was it that drew a young electrical engineer into this new world? Searching for the elusive Robert Paul is also exploring the fascination with new technology that had launched telegraphy and electric lighting, and would soon bring X-rays and electric tramcars and automobiles, before moving pictures emerged as a new collective pastime for the whole world—the pleasure of spending time, traveling far and wide, without leaving one's seat in a darkened room.

The Place of Moving Pictures

Where were moving pictures invented? Traditionally, the story of cinema's origins focuses on Thomas Edison in New Jersey, with the Kinetoscope peep-show viewer that W. K. L. Dickson developed for him, and on the Lumière brothers in Lyon, with their Cinématographe, inspired by the success of Edison's invention. Certainly there were many other inventors working on moving pictures in half a dozen countries, often with promising devices, but these two became the most widely known and commercially successful in the early years of the new medium.

A third location has been consistently ignored: London. And with it, the engineer turned producer who was much more closely involved in the process of invention than Edison, and who contributed directly over fifteen years to steering the novelty toward what became cinema. Robert Paul, scientific instrument maker and onetime acknowledged "father" of the British film industry, has been steadily demoted in the widely accepted "story of cinema," and even branded by some contemporary historians of early cinema as an opportunist and deceitful self-promoter.[1]

Quite apart from such charges, it is clear that Paul has failed to register in popular consciousness as one of the founding fathers. No streets or universities are named after him, as they are after Edison and the Lumières. Only two small wall plaques, erected during the 1995–1996 centenary of cinema celebrations in Britain, publicly acknowledge his role, one on the site of his workshop in Hatton Garden in central London, and the other on a house he built, next to his studio and factory, in the north London suburb of Muswell Hill. Paul wanted his pioneering status to be known among a professional audience — donating historic equipment to what would become Britain's Science Museum and supplying information about his early achievement — but was otherwise unconcerned with public recognition and deliberately avoided the limelight. However, "animated photography" was news in 1896, and the popular *Strand* magazine ran an illustrated article about him filming the annual Derby horserace at Epsom Downs, one of Britain's major national traditions, in June of that year.

Among film historians, Paul's status has long been recognized, although often in an oblique manner. As far back as Terry Ramsaye's pioneering *A Million and One Nights*, published in 1926, Paul received a whole chapter to himself, intriguingly entitled "Paul and 'The Time Machine.'"[2] This popularized the story of how Paul became "accidentally" involved in moving pictures and how his early interest led to a speculative project for a "time machine" inspired by H. G. Wells's famous story, then newly published. As well as having a shrewd journalist's eye for a good story, Ramsaye was an assiduous researcher, basing his narrative on correspondence with both Paul and Wells. While Ramsaye's account of this almost legendary episode seems generally reliable, and was indeed confirmed by both men near the end of their lives, it severely underestimates the extent of their foresight. Paul would quickly become perhaps the key architect of the new medium of projected narrative pictures, while almost as rapidly, Wells imagined a world in thrall to such imagery in his early scientific romance *When the Sleeper Wakes*.[3]

The pioneer historian of British cinema, Rachael Low, identified Paul as one of the founders of the British film industry, but credited this to his "business and scientific abilities [rather] than to any artistic gifts," describing his films, somewhat bizarrely, as "characterized by a lack of either taste or appreciation of the larger possibilities of the cinema."[4] But even as Low delivered this strange verdict, her French contemporary Georges Sadoul cited evidence of Paul's fertile invention and influence on French filmmaking. Forty years later, John Barnes's minutely detailed study *The Beginnings of Cinema in England* was highly respectful — to the extent that other historians have sometimes considered it unduly biased in favor of Paul, especially in discussing his short-lived partnership with Birt Acres.[5]

So is the issue merely one of public perception—that Paul is not a household name in Britain as Edison and Lumière are, in their native countries as well as internationally? Certainly there is a larger question about how the British perceive—or rather, ignore—their place in what Ramsaye called "the world of the screen." For the first half of the twentieth century, British film history was wedded to the myth of an unacknowledged pioneer, William Friese-Greene, long regarded as the "true inventor" of moving pictures, whose claim was believed to have been ignored by Edison and all who came after. The subsequent discovery that Friese-Greene had not achieved what was erroneously claimed on his behalf seems to have left a sense of embarrassment that his position had been so long championed.[6] If inventing cinema is seen as a race, then the British apparently either cheated at the start or quickly lost their early lead. In either case, a contentious first chapter may have contributed to the distinctly ambivalent national attitude toward cinema that continues up to the present.[7]

But of course the many scattered inventions, commercial strategies, and social currents that came together as "cinema" in the early years of the last century cannot plausibly be considered a race—even if competition was, and has remained, a vital dynamic. Bringing Paul out of the shadows as another "neglected pioneer" would be a questionable aim, not only because he has never been forgotten, but also because such moves evoke the dubious historiography of "great men." It is surely significant that so little attention has hitherto been paid to establishing the full range of what he contributed to the nascent industry, and art, of cinema. Yet the deeper problem with Paul is deciding just *what* he represents: an inventor, entrepreneur, pioneer filmmaker, producer, studio head—or an engineer temporarily beguiled by this new apparatus, an opportunist who claimed credit for others' inventions, a businessman who dabbled briefly in film? During its relatively short academic career, film studies has been preoccupied with identifying creativity in an otherwise industrial process, and has invested heavily in the idea of the director as the prime authorial figure in cinema. This approach has little obvious application to early filmmaking, where many roles were undertaken by the same individuals, within organizational structures that are still little understood and wholly undocumented. Nonetheless, Paul has been routinely identified as a "producer," while others known to have worked for him are considered "directors."

Ramsaye, Low, and Barnes—pioneers themselves in writing the history of the medium's early development—realized that filmmaking could not be approached as a linear or unitary process, and devised their own structures to deal with its different facets. Later historians of this period, such as Charles Musser, Barry Anthony, and Richard Brown, have laid increasing emphasis on the business dimension, showing how impersonal factors interacted with the choices and decisions of individuals.[8] This contextual historiography, in contrast to earlier approaches based primarily on surviving film "texts"—from which authors and spectators were hypothesized—certainly provides a better framework for assessing the career of a figure such as Paul. But it too requires considerable guesswork, since, in the case of Paul, there are no known company records and few personal papers. And it leaves open the issue of how to assess his output of films, of which less than a tenth are currently known to exist.

Writing a conventional biography of Paul would be impossible, given the paucity of sources. And writing about him only as a prototype film auteur or entrepreneur is problematic. What I have attempted is to collect as much information about Paul as seems possible, and to set the early moving-picture business in the wider context it inhabited—and in which cinema has always existed.[9] Film scholars have often bemoaned how little attention is paid to cinema by social, economic, and other types of historian, who usually assign it a minor role as symptom or reflection of larger processes. Certainly, moving pictures became ubiquitous within a few years in the late 1890s. Ubiquitous, and soon pervasive, but by no means dominant. They were a novelty in an age of novelties—a part, and often a relatively small one, of most producers' and middlemen's business activity, as they were of consumers' leisure. No one could have predicted that they would emerge from the mass of competing attractions to become, along with the phonograph and the radio, the cornerstone of a new social and industrial order—no one, that is, except Wells in the extraordinary blueprint offered to his awakened Sleeper (of which more later).

What we can find in the earliest period are different ways of relating this novelty to the world around it, which may help us understand why and how it became so pervasive. And tracking the elusive figure of Robert Paul provides a thread, showing how chance and luck created opportunity, which Paul enthusiastically seized, leading to a decade of extraordinary creativity, which richly rewards close attention and deserves wider recognition, especially in Britain. Nor do we have to accept, as Paul himself implied, that instrument making was a completely different world. Trying to encompass his career as a whole reveals the extent to which moving pictures were a *scientific* novelty, and so a natural extension of Paul's lifelong concern. Yet most elusive of all remains the audience, his audiences, which we can only glimpse in occasional asides and early re-views, or infer from the exponential growth of the business.[10] If, as Luke McKernan has claimed, it was the early audience that "made" cinema, as much as its literal manufacturers, then Paul may help us discover more about these — our ancestors.[11]

CHAPTER 1

Getting into the Picture Business

One of the most famous anecdotes in early cinema history involves Robert Paul. It seems too improbable to be true, but this apparently is what happened. Two mysterious businessmen, of Greek origin but from New York, call on Paul at his Hatton Garden engineering business in 1894 and invite him to replicate or copy the latest mechanical marvel, Thomas Edison's Kinetoscope. To his surprise, he finds there is no legal obstacle and proceeds to do so, initially for the Greeks, then for himself and other customers in Britain and beyond.[1] And so a career in "animated photography" was launched, with London becoming one of the capital cities of the new era.

But what was the Kinetoscope? A primitive forerunner of cinema proper, or a magic box that first revealed the potential of moving pictures? During much of the twentieth century, it seemed to belong firmly to the primitive category, part of the vast lumber-room of "pre-cinema" devices that preceded the real thing. But now that the "real thing" has splintered into so many different forms of delivery, from giant IMAX screens to mobile phone displays, we are free look afresh at the Kinetoscope and realize what an important step it marked. For two years, till the end of 1895, it was hailed as the marvel of the age. It was not yet the *forerunner* of projected moving pictures, which didn't exist, except fitfully in the workshops and dreams of a few scattered experimenters. Rather, it *was* moving pictures, and it remained the reference point when projection first appeared.

The Magic Box

The Kinetoscope resulted from some five years of trial and error by Thomas Edison and members of his staff at West Orange, New Jersey. Technically and conceptually, their experiments built upon a series of sci-

2. Early publicity illustration of Edison's Kinetoscope viewer, showing its continuous filmstrip.

entific discoveries about vision made early in the century, and the family of popular "optical toys" that followed.² But for Edison, its immediate origins lay in the astonishing success of his Phonograph, which had aroused worldwide admiration for "the wizard" in 1877. The first versions of the Phonograph used tinfoil-covered cylinders and produced very limited sound quality, adequate for speech recording but not for music; it was clearly the *idea* of being able to capture and store sound, the potential of this new magical instrument—the second modern time-shifting device to appear after photography—that mattered. Nonetheless, even before Edison introduced a "perfected" version in 1888, the Phonograph had captured the world's imagination and inspired ideas for its use that ranged from attaching it to the front door to take callers' messages, or combining it with the newly invented telephone, to ambitious schemes for long-distance medical diagnosis, and even, as an extension of Victorian "memorial culture," fantasies of communicating with the dead.[3]

In 1888, after a meeting with the photographer Eadweard Muybridge, who was showing sequential photographs of figures in motion using a projection version of the phenakistiscope optical toy, Edison announced that he planned "to do for the eye what the Phonograph did for the ear," in other words to offer continuous recording.[4] He set his assistants to work on a device that would use the same cylinder format, with a spiral of images wrapped around it, although the search for a workable system would eventually lead to a flexible band or strip, passing over rollers. Most of the actual development was led by the French-born Scot William Dickson, who adopted a new transparent material, celluloid, that offered strength and flexibility, while accepting photographic emulsion.[5] And so the basic format of the motion-picture "film" was born. Indeed, Dickson's decisions on the strip's width and perforated holes on either side to allow the film to be moved by sprocket wheels became standard and universal for a century, as did the principle of creating sequential images and changing these sufficiently fast for spectators to experience the illusion of a continuous moving reality.

But although the Phonograph had led Edison to the Kinetoscope, there was another midcentury invention that arguably did as much to create an appetite for lifelike reproduction of the visible world. The stereoscope tricked the eye in another way, by using two photographic images taken from slightly separated viewpoints, which when seen through a special viewer offered the illusion of a "solid," three-dimensional image. Developed from Charles Wheatstone's initial study of stereopsis in normal vision in 1838, the stereoscope would, in various forms, become the first widely available instrument of the nineteenth-century media revolution during the 1860s and 1870s. Many hundreds of thousands of viewers were produced according to the Brewster or Bates-Holmes pattern, and the total number of stereograph cards in circulation must have run into millions, ranging from famous places and sights to works of art and biblical scenes, and from exotic peoples pictured by explorers to erotic poses for private consumption[6] The illusion only works if the viewing device is held close to both eyes, providing an intensely *personal* experience in which normal perception is suspended; the user's immediate surroundings are replaced by a specially arranged image that appears "super-real."

One of the earliest enthusiasts for the stereoscope was the American poet and essayist Oliver Wendell Holmes, who wrote evocatively of the experience in 1859:

> The mind feels its way into the very depths of the picture. The scraggy branches of a tree in the foreground run out at us as if they would scratch our eyes out. The elbow of a figure stands forth so as to make us almost uncomfortable. Then there is such a frightful amount of detail, that we have the same sense of infinite complexity which Nature gives us.[7]

Not everyone shared this enthusiasm, especially for the sheer scale of its adoption. The poet Charles Baudelaire, skeptical about the impact of new technologies on art, while also aware of the bond that is formed with toys in childhood, wrote contemptuously of "thousands of greedy pairs of eyes bent to the stereoscope's openings as to the skylight to infinity."[8] But the meteoric success of the stereoscope as a new way of seeing and a device that redefined both the spectator—who became a kind of voyeur—and the spectacle, may be a better pointer than the phonograph toward the appetite for visual novelty and entertainment that moving pictures would satisfy.[9]

Another, older visual device had simultaneously been undergoing spectacular changes. The magic lantern dated back to the seventeenth century, but the nineteenth century brought dramatic improvements in what had hitherto been largely a domestic entertainment. New forms of illumination, especially "limelight," enabled lanterns to show large, bright images before vastly increased audiences. From the 1840s onward, photographic slides greatly expanded the repertoire available to lanternists, making possible "travelogues" and narratives using "life models"—both genres that also flourished in stereoscopic form, and would continue into the moving-picture era. Double and triple lanterns became increasingly popular, allowing lanternists to project elaborate sequences of "dissolving views," which implied the passing of time. And "slipping slides," with moving parts, became increasingly ingenious, with gear-driven kaleidoscopic "chromatropes" and mechanisms making possible the illusion of anecdotal movements—such as a "man swallowing rats." Cumulatively, by the closing decades of the century these developments had made the newly renamed "optical lantern" (or "stereopticon" in the United States) a versatile device, capable of creating many kinds of spectacle and offering an immersive experience to large audiences. Projected moving pictures too would depend upon the lantern, with an extended period when slides and moving pictures were projected alternately from the same dual-purpose device.[10]

If the phenakistiscope and its many descendants offered forms of repeatable *movement*, then the enhanced lantern and stereoscope created a popular taste for different versions of "presence"—of the illusion appearing somewhere other than the place of viewing. There were indeed many attempts to create a stereoscopic form of lantern projection, just as there was early optimism that moving pictures could be taken and shown stereoscopically.[11] Although neither proved possible within the parameters of "normal" turn-of-the-century spectatorship, "animated photography" was soon found to provide an attractive and practical hybrid experience.

Thanks to the established popularity of the stereoscope, the Kinetoscope could be developed as a viewing device that required the user to look into a lens to see its moving images. As a technical showpiece, the Kinetoscope featured Edison's even more famous invention, the incandescent electric light, and was one of the first electrically powered machines to be widely displayed. These novel features compensated for an image that, compared with the stereoscope's immersive character, was small, needing to be viewed through a magnifying lens, and rather dim (a rotating shutter that segmented the continuously moving film strip blocked much of the light from reaching the viewer). As an economic venture, it was also one of the earliest coin-operated entertainment device, with a business model—a single viewer paying for a brief glimpse of "living pictures"—that both financed the expensive machine and tapped into a popular wave of mechanization, as automatic vending and gambling machines began to appear on both sides of the Atlantic.

The first such machine to be deployed publicly on a large scale was a postcard-vending machine devised by Percival Everitt in 1883. Over a hundred were installed around London, and the US patent was acquired by Thomas Adams to sell his Tutti-Fruiti chewing gum across the New York subway system.[12] In the 1890s, Gustav Schultze and Charles Fey began producing automatic gambling machines in San Francisco. Soon Sittman and Pitt's similar machines were common in New York bars, suggesting a forerunner of the kinetoscope parlor, with its brief bursts of kinetic excitement. In Germany in 1892, Ottomar Anschütz's electric "Schnellseher"—literally "Rapid Viewer"—showed images on a celluloid disk and was also coin-operated. All of the new American devices used nickels, which may well have guided the Edison laboratory's decision that the Kinetoscope should take the same coin. And the link between early moving pictures and automated candy sales in particular would continue. The British pioneer (and Paul's temporary associate) Birt Acres was hired in 1895 by the German chocolate company Ludwig Stollwerck to make films for its Kinetoscope parlors; while in 1897 another American gum company would introduce animated figures as an extra attraction on its vending machines, as if acknowledging Edison's success.

The Kinetoscope had had a fitful history of development, as Edison's own interest in it fluctuated over five years while he struggled with other, potentially more lucrative inventions. Delayed beyond its intended launch at the 1893 World's Columbian Exposition in Chicago, where the application of electricity was a major theme, it had its first public demonstration in May 1893 at the Brooklyn Academy of Arts and Sciences. Early descriptions of the chest-high wooden cabinet that housed (and hid) the mechanism compared it to "a medium-sized refrigerator" or simply "a nice piece of furniture," with a peephole or viewing aperture at the top.[13] Inside the cabinet, a continuous strip of celluloid passed over a series of rollers at approximately 40 frames per second, the viewer seeing this by means of an electric light powered by batteries and made intermittent by a rotating perforated wheel. Although the Kinetoscope undoubtedly gave an impressive illusion of natural movement, it did not use the intermittent strip-movement principle common to all later projectors and motion-picture cameras.

Like other devices that exploited scientific principles and new technologies, it still enjoyed an ambiguous status, part "demonstration" and part novelty, waiting to see if there was a market. Edison admitted his doubts that it had a commercial future in some early interviews,[14] and the fact that he did not seek patent protection beyond the United States has often been attributed to this. Another explanation, however, is that, having spent several years and much staff time on it, he realized some of its features would not be patentable abroad, where other moving-picture devices already used perforated translucent strips and intermittent illumination.[15]

Whatever the reservations about its potential or parentage, the kinetoscope made its commercial debut in a special "parlor," or arcade, on Broadway on 14 April 1894. With ten machines, each offering a different subject, the Holland Brothers' venture proved an immediate success, leading them to open similar parlors in Chicago and San Francisco within six weeks. Unlike much of the subsequent history of moving pictures, the economics of this first phase of exhibition are largely known. The Hollands paid $300 for each machine to the Kinetoscope Company, which in turn bought them for $200 from Edison. Against this relatively high outlay, and the high cost of viewing—5 cents for less than 30 seconds—they apparently grossed an average of $1,400 per month for the first year, which translates to an average of a thousand customers per day.[16] Other companies sprang up to open kinetoscope shows in many cities around the world—one of which was London, where the first parlor opened on Oxford Street on 18 October. Before long, there would be more than six venues in London, including one operated by Paul's Greek-American customers, George Georgiades and George Trajidis.[17]

What exactly was drawing such crowds, and who was it attracting? No doubt, at the beginning it was literally novelty—to be able to see "life" reproduced by the addition of movement to familiar "still" photography. Edison had worked hard to establish his name as a brand, promising endless new marvels, and was already associated with the lifelike recording of sound.[18] But the choice of subjects photographed for the first kinetoscope loops was also shrewd. Six of the initial ten were variety acts of the kind that might be seen on the vaudeville or music-hall stage (two featuring Ena Bertholdi, a British-born contortionist; one of the strongman Eugen Sandow posing; a "Highland dance"; and trapeze and wrestling acts). Two, *Blacksmiths* and *Horse Shoeing*, were genre scenes, already common subjects for photographs or prints. *Barber Shop* was a prototype comedy, a cartoon subject brought to life. And *Cock Fight* was a sporting scene, albeit of a sport already banned in most American states and in Britain. All had been filmed in the specially constructed studio at Edison's factory in Orange, nicknamed the "Black Maria," after the armored vehicle used to transport prisoners, because of its rather sinister shape and black tarpaper covering. Their main purpose was of course to demonstrate movement, but they also offered a varied bill of attractions, based on the vaudeville stage, while also gesturing in other directions.

Discovering who attended the kinetoscope parlors is inevitably more speculative. Anecdotes recorded in newspapers during 1894 and 1895 turned on how rapidly word of the wonder had spread, and the impact it had on the unsuspecting. A Syracuse, New Jersey, newspaper declared in December 1894, "Everyone knows that the kinetoscope is the device by which a prize-fight, a family row, skirt dancer . . . can be reproduced pictorially."[19] And a March 1895 account of a first encounter already patronized the inexperienced:

The expression on the customer's face undergoes a swift change. . . . He gazes at the picture in rapt amazement, as if he expected the figures in it to speak. Before he recovers from his surprise the vivid scene is blotted out in a snap.

He lifts his astonished eyes from the picture, and looking up exclaims: "By gosh, I've allus heard tell that them livin pictures was great."[20]

Accounts also vied in conveying the lifelikeness of the representations, as in the anecdote of a boxer watching the Corbett-Courtney fight on a Kinetoscope on Broadway and bragging that, on the basis of what he'd seen, he "could punch [Corbett's] head off."[21]

One of the few available photographs of a parlor with people present, from San Francisco, shows two women at the back of the row of machines, where three men (probably the proprietors) pose stiffly. And in a promotional illustration of the period, published in New York, a fashionable lady is placed prominently in the foreground. An advertisement from Bradford in December 1894 spoke of

expressions of astonishment at the wonders which the instruments reveal. It is undoubtedly the duty of everyone who is interested in the progress of events or who wishes to be thoroughly "up-to-date" to visit this scene of attraction. Not that the Kinetoscope, any more than the phonograph, is exclusively of scientific interest. It is also a means of entertainment.[22]

All of this circumstantial evidence points to the Kinetoscope's attracting, or at least intending to attract, a mixed clientele of men and women—with some emphasis on its suitability for unaccompanied women—and also underlines the "improving" or educational dimension of displays of new technology at this time.

Edison's original large Kinetograph camera required a stable mounting and electrical supply, with strongly lit subjects filmed against a background of black drapes to maximize contrast, producing "the singular distinctness of the kineto strips" described by Dickson, which ensured they would remain legible when viewed in the Kinetoscope.[23] These factors may not have wholly determined what Dickson and his colleagues chose to film. But their choices *did* serve to attract the first audiences. And what happened next provided something like a rehearsal of the course of cinema itself, all compressed into less than eighteen months. Soon after the first variety subjects came multipart pictures, with a boxing match specially staged for the Kinetograph in June, so that it could be shown, round by round, on a row of Kinetoscopes; next, a range of dancers offering ever more alluring images of the scantily dressed female body; and by December, the grand finale of a full-scale Broadway musical, Hoyt's *The Milk White Flag*, with "thirty four persons in costume." During the next year, the repertoire would broaden even more dramatically, as independent manufacturers entered the field outside the United States and also started producing films for their Kinetoscopes. The largest body of these films still extant is by Paul and Acres, but others were active as well, even if their efforts are now lost.

Mainstream cinema history has traditionally assigned the Kinetoscope a minor and transient role in the development of moving pictures, but this is to fall into the trap of thinking that media evolve sequentially, with each replacing its predecessor. We should not forget that almost all the films first shown on screens, other than those of the Lumières, were originally made for Kinetoscope use, and that almost all of the first generation of moving-picture makers and spectators began with the Kinetoscope. Less obviously, the device has left its own cultural trace, in such classics as Wells's *The Time Machine* and Marcel Proust's *A la recherche du temps perdu* and in many lesser known works.[24] Nor did it simply disappear in early 1896, when the first screen projection began in Europe. Again the traces are scant, but there is evidence that Kinetoscopes continued to be operated until the end of the decade.[25] And its early rival, the robust Mutoscope, using single cards instead of a filmstrip, would continue to provide popular entertainment well beyond the middle of the twentieth century.[26]

Meanwhile in London

Exhibitions of the Kinetoscope did more than attract crowds. They attracted imitators, who either copied or elaborated upon Edison's design. The arrival of the Kinetoscope in Paris prompted Antoine Lumière to invite his sons, Louis and Auguste, to consider how its achievement might be improved, which led to their creating the Cinématographe in 1895. In England, the Kinetoscope was soon duplicated by Robert Paul.

Direct imitation is perhaps the most neutral term for this act of copying. Paul's short-term associate Birt Acres would later, in the margins of Frederick Talbot's 1912 book *Moving Pictures*, deem it "pure piracy."[27] But this is to run ahead of the fateful events of mid-1894, which are not in dispute, even if Paul's ethics are. The first account of Paul's meeting with the two Greeks appeared in Talbot's book, which cast the visitors as previously a greengrocer and a toymaker.[28] A more detailed account by Terry Ramsaye, published in 1926, introduced the Greek-born businessman

George Georgiades, who was said to have already exhibited the Kineto-scope in New York before arriving in London with his partner George Tra-jidis (or Tragidis) to set up a similar show, some time after Maguire and Baucus established theirs on 18 October 1894. The basic problem facing exhibitors was the price of £70 being charged in Britain for the Edison device, to which had to be added transport of the heavy battery-operated machines from New York and the purchase of new film loops.[29] Unsur-prisingly, Georgiades and Tragides were looking for ways to economize in expanding their existing exhibition in Queen Street, off Cheapside, which had been under way for some time in December.[30] Although the Kinetoscope had proved the most successful Edison "marvel" brought to market since the Phonograph, the investment it required from exhibitors was significantly higher — and unlike the Phonograph, it served only one customer at a time.

Ramsaye's colorful account of how the pair came to proposition Paul has been widely repeated, even if its details are impossible to verify:

> Georgiades and Trajedis [*sic*] fancied a certain brand of tobacco in their cigarettes, which made them familiars in the cigar[ette?] establishment of John Melachrino, the same whose name can still be observed these many years later on a machine-made product of a similar character.
>
> To their countryman, Melachrino, the Greeks confided their quest. The course of motion picture destiny had now become as thinly fanciful as the curling smoke of a cigarette. Melachrino in turn recalled an English customer for that same fragrant blending of the weed, one Henry Short. Now Short had gossiped over the counter of the skill of his old friend Robert W. Paul, maker of scientific instruments of precision. At Mela-chrino's the Greeks waited for Short and he took them to Paul with their mechanical problem.[31]

According to this and all other accounts, Paul's first response was that he could not infringe Edison's patent, but when he discovered that the patent did not cover Britain or any other territory outside the United States, he accepted the commission. Indeed, he went on to become prob-ably the largest single manufacturer of Kinetoscopes, supplying not only Georgiades and Tragides, but other exhibitors around Britain, his future rival Charles Pathé in France, and customers as far afield as Japan, as well as exhibiting them on his own account.

But how reliable is Ramsaye's account? A tobacco shop certainly existed on Cockspur Street, near Trafalgar Square, and Short could easily have recommended the two businessmen to Paul. The only significant variation on the story seems to be in a manuscript by the pioneer film historian and collector Will Day, who wrote in 1932:

> These two Greek gentlemen, being very astute business men, soon made enquiries of various firms regarding estimates for manufacturing similar machines and films in England. Their choice for the placing of orders fell upon Mr R. W. Paul, who was recommended to them as a very clever maker of scientific and optical instruments.[32]

But whether or not Georgiades and Tragides considered other possible manufacturers for what was effectively a new kind of apparatus, it was apparently their unannounced arrival that lit the fuse that led to Paul's sustained involvement in cinema for the next fifteen years. Like the Lumières, he seems to have had no previous interest in solving the moving-picture problem. Paul's agreement to make facsimile Kinetoscopes would lead to legal action, although not against him. Recent research by Brown and Anthony has revealed a previously unknown case brought against Georgiades and Tragides in January 1895, accusing them of "passing off" the Kinetoscopes made by Paul as genuine Edison machines.[33] The issue in this case was not whether the English productions infringed Edison's patent, which he had indeed not extended internationally, but whether the machines were being fraudulently represented to potential buyers. The court ordered Georgiades and Tragedis, and their associate James Hough, to desist from representing their machines as Edison's, and they formally dissolved their partnership as the "American Kinetoscope Company" in early February.[34]

Many questions arise from the story of the 1894 "commission," and few can be answered with any certainty. The various roles played by Henry Short in 1894–1896 seem particularly significant, and it transpires that his father, Thomas Short, was in fact an instrument maker, already based in Hatton Garden, with his company located at no. 40, close to Paul's premises at 44.[35] Within this relatively small world, it seems likely that Paul and Henry Short could have known each other for some years before 1894 — and of course that Short would know where Paul was located. Short would make another fateful connection early in 1895, introducing Paul to Birt Acres, when photographic expertise was needed to create a moving-picture camera. Short is credited in various sources as having photographed a series of subjects for Acres in Germany that June.[36] Acres, however, in the 1912 notes quoted earlier, asserts he was *his* "paid assistant Henry W. Short" when he appeared in a short film, known as *Incident outside Clovelly Cottage*, made at Acres's house in Barnet, probably in March 1895. This description suggests that by 1912, when Acres felt deeply wounded by Talbot's account of his relations with Paul, he might have wanted to suppress any suggestion of collaboration with Short as anything other than an employer. Early in 1896, Short patented his own form of intermittent mechanism for a film camera or projector, although he seems not to have pursued this, undertaking instead a pioneering expedition for Paul to film the series *A Tour in Spain and Portugal*, and in the following year, a similar series of touristic scenes in Egypt. However, in 1898 he also devised and patented the Filoscope, a form of miniature photographic flipbook in a metal case, which used a dozen of Paul's films for its subjects, suggesting that he had an active interest in the business side of the emerging cluster of moving-image technologies.[37]

A possible further explanation of Short's role as "foreign" cameraman for both Acres and Paul is offered by Irene Codd, who wrote that "a young engineer friend had come to Paul and said 'I can't stand father any longer.'"[38] Motivated, she claims, by "his own youthful experience" of his father opposing his choice of career, Paul asked if the young engineer "would like to go abroad and take films for him." And thus "he found himself sent abroad to Venice to photograph the beauties of that city." Now, although there is no record in Paul's lists of a film taken in Venice, there is one attributed to Acres in 1895, *Venice: The Rialto Bridge*—also known as *Venedig: Die Rialto Brücke*, because it was in fact filmed, not in Italy, but at the Italian exhibition in Hamburg, where indeed Short was working with or for Acres in June. Given Codd's friendship with the Pauls, it seems at least possible that this jumbled account is based on some genuine knowledge of what linked Short and Paul—and perhaps why Acres later repudiated him. At any rate, Short was undoubtedly involved in the two vital introductions that determined Paul's entry into movies.

Paul apparently had access to one of Georgiades and Trajidis's Kinetoscopes as he set about making his own.[39] Manufacturing such large machines must have required a step-change in his existing business, hitherto devoted to making relatively small-scale instruments. It would have required new staff, enlarged workshop premises, and considerable financial risk—which would of course explain why he insisted on a contract with Georgiades and Trajidis. None of these implications for his expanded business can be properly explored with the scant information we have, but we may suppose that manufacturing Kinetoscopes brought Paul into contact with new business and social contacts, outside the world of electrical engineering—a process that would accelerate in the next few years.[40] In any case, Paul evidently discovered how to produce Kinetoscopes in bulk and at a profit, and must also have become sufficiently aware of how they were shown to become an exhibitor in his own right. His showings at Earls Court exhibition center in West London, during the *Empire of India* exhibition, were undertaken, he wrote in a 1936 paper presented to the British Kinematograph Society, "in conjunction with business friends," although these have never been identified.[41]

Paul's Kinetoscope manufacturing business was flourishing, but it depended vitally on the continued supply of films, or "subjects," all of which came from the Edison Manufacturing Company, with its still unique Kinetograph camera. Having lost control of the foreign market for Kinetoscope machines early in 1895, Edison's managers attempted to reassert control by cutting off the sale of subjects to unauthorized operators. In a business that depended on regularly refreshing the repertoire, the only solution was devise a camera—without the benefit of access to Edison's original.

Once again, Henry Short made the introduction, putting Paul in touch with the American-born photographer Birt Acres, who already had an interest in moving pictures, presumably thinking that these two could help each other.[42] Most of what happened next would be disputed by Acres, who believed Paul had deliberately minimized his contribution to their joint work and would eventually claim that he alone was responsible for a process of invention and film production that dated back to early 1895. Two later histories complicated the already disputed story, one published before Acres's death in 1918 and the other, Ramsaye's, in 1926, which attempted to establish the truth. Ramsaye's account was widely accepted until the 1990s; since then Acres's has gained ground, casting doubt on Paul's credibility and character in some quarters.[43]

Where to begin an assessment of the rival claims? In his last account, Paul explained his move into production as due to "Edison agents" refusing "supplies of films ... to users of my Kinetoscopes." He continued:

In Birt Acres I found a photographer willing to take up the photography and processing, provided I could supply him with the necessary plant, which I did early in 1895.[44]

Paul had written to Ramsaye a decade earlier, apparently after referring to documents, saying, "I find that on March 29, 1895, I made an agreement with Birt Acres for him to take films for the Kinetoscope with a camera made by me."[45] This may be so—Acres never saw it in print—but it does not mean this is when they began to collaborate. March 29 is in fact the date of a surviving letter from Paul to Edison, which enclosed sample frames of a film shot by Paul and/or Acres—the film taken in front of Acres's house, with Short appearing. The partnership had actually begun on 4 February, with work on the camera starting the following day. A simple explanation of the coinciding dates would be that writing to Edison to propose that they exchange films or collaborate on production required a formal agreement with Acres to cover what had previously been an informal arrangement.

Acres's rebuttal of Paul's account, which separated the mechanical and the photographic, began in an article for the *Amateur Photographer*, in October 1896, that referred pointedly to "various people [who] claim to be the inventor of animated photographs some of them without the slightest shadow of a claim to the title."[46] He then set out the claim that

> in the spring of 1894 I designed the apparatus with which (with modifications) I have taken the pictures many of which are now so well known.
>
> Without claiming to be the originator of animated photography I do claim to have been the first to have made a portable apparatus which successfully took photographs of ordinary scenes of everyday life. In March 1895, I took a successful photograph of the Oxford and Cambridge Boat Race, and in the same year I photographed the Derby.

Acres had by then constructed his own chronology, which completely omitted Paul, making no reference to their partnership, while referring to some of the films generally attributed jointly to them—essentially the "scenes of everyday life" that began with Henley Regatta and continued with *The Boat Race* and *Rough Sea at Dover*.[47]

How did Paul explain the partnership? In the 1936 paper he described developing the camera as a collaborative process:

> We had no information to guide us in making a camera, but I worked out an idea due to Acres. The film was drawn, under slight tension, from an upper spool, past the light opening, or gate, to another spool below. A clamping plate, intermittently actuated by a cam, held the film stationary in the gate during each exposure. A shutter, whose opening synchronized with the cam, revolved between the lens and the gate. In our first trial, we failed to get a picture on Blackfriars Bridge, only because we forgot, in our excitement, to attach the lens.

In his 1912 notes, written after reading Talbot's account, which referred to "the rough drawing Acres had made of his bromide printing process" and

3. Birt Acres filming the 1895 Derby with the Paul-Acres camera. Henry Short is believed to be the figure on the extreme left, wearing a cap. National Museum of Science and Industry.

how this "gave birth ... to an entirely different project," namely the intermittent mechanism, Acres wrote on his copy (using the third person):

> This was the work of Acres, not Paul. At the time Acres showed his design, Paul had not the faintest idea that it was necessary to arrest the film for exposure.[48]

Earlier in the same marginal notes, after dubbing Paul's undertaking the Kinetoscope commission "Pure Piracy!," he wrote:

> I say that Paul never made or had any idea how to make a film for the Kinetoscope until I made them for him. Even then I did not allow him at my works as I knew that any ordinary mechanic could copy the Kinetoscope, but making the pictures was a vastly different thing.

Another contemporary, Will Day, gave a different reason for why the camera development work came to be done at Barnet:

> The first dark room set up by Mr. R. W. Paul was at Saffron Hill, but as space was a consideration at these premises, it was found more practical

and satisfactory for Mr Bert [*sic*] Acres to produce the pictures and carry out development etc. at Barnet.[49]

In 1936 Paul showed as part of his presentation to the British Kinematograph Society "a bit of Kinetoscope film taken during a trial of our first camera in February 1895," and spoke of "we" in relation to the filming of *Rough Sea at Dover* and *An Engineer's Smithy*.

Two key issues seem to be in dispute: was the making of the first camera collaborative, as Paul stated, or solely Acres's work, as he claimed? And, whatever the credit for the camera, was the making of the films produced with it between February and June in any way collaborative or solely Acres's responsibility? One important consideration not mentioned by either party is the fact that their agreement, originally envisaged as lasting ten years, had been ended by Paul, either in June, after learning that Acres had patented his Kinetic camera on 27 May, based on their joint undertaking, or in July, when Acres refused to make any more films without funding. Acres may well have felt justified in claiming the camera as largely due to his work—although Paul apparently considered this to be breaking their agreement—and he maintained that Paul had reneged on an undertaking to buy a minimum quantity of the films he had made.[50] Both men, in any case, would proceed to claim and show the films dating from their partnership, although Acres continued to regard them as "*my pictures*" and regarded the Talbot book as "untrue," "garbled," and

intentionally mis-stated for the purpose of representing Paul (*the copier of other peoples' work*) as the Pioneer of Animated Photography.

Fourteen years later, Ramsaye wrote diplomatically that these "enthusiastic patriotic claims of priority" came from Talbot rather than Paul, and quoted from Paul's "conservative and precise" response to his questions to set the record straight. But Acres had died in 1918, still bitter, and believing that Paul had built a pioneer reputation at his expense.

Since Paul referred to these disagreements, it seems likely that he knew the extent of Acres's animosity. And indeed, from earlier correspondence in the electrical press, it is clear that Paul was no stranger to controversy over credit and rights.[51] Yet perhaps the strangest point to emerge from Ramsaye's account is his quotation of another part of Paul's letter, which seems to confirm Acres's claim that Paul had not understood the need to "arrest the movement of the film," although he makes the admission in relation to developing a projector:

Early in 1895 I had not realized the necessity for intermittent movement of the film, but it is certain that I did so by October of that year; this of course refers to projection, and is the stranger as I had already made several cameras with intermittent motion.[52]

One interpretation of this might be that Paul had indeed been more remote from the actual development of the camera than his other accounts suggested. At the very least it confirms that Paul had not approached the film process in the same spirit as the Lumières, who saw camera and projector as essentially the same mechanism — and in their case, literally the same machine. The only other insight we have into Paul's side of the disputed story is a letter he sent to the German pioneer Oskar Messter, in response to an inquiry in 1932. Messter had his own reasons for checking details and dates for the early period of invention, which will be discussed in chapter 11, but the surviving letters offer an early draft of what Paul would later say to the BKS. In the first letter, he confirmed that he made six Kinetoscopes for Georgiades and Trajidis in 1894 "and later about 60 more."[53] When Maguire and Baucus cut off the supply of films to buyers of his machines, he began the development of a camera:

In January 1895 I became acquainted with Mr Birt Acres, a traveling representative for Messrs Elliott and Son, makers of plates and paper for photography, of Barnet. He informed me that Mr Blair, of Foots Cray, Kent, was making photographic film and could cut it to the required width for the Kinetoscope. Neither of us had any experience of "Cinematography" but Mr Acres was a photographer. He proposed, and it was agreed, that I should make a perforator, camera and printer, which he would use to make pictures exclusively for me. On March 29th 1895 a formal agreement was signed to that effect.

Later in the same letter, in response to Messter's quotation from Henry Hopwood "that Mr Byrd [*sic*] Acres had taken kinematographic photos already in March 1895," Paul wrote emphatically:

With regard to Mr Birt Acres' early work, I am sure that he had no experiments in Moving Pictures before he had the camera which I designed and made in 1895.[54]

If we accept that the "Paul-Acres camera," as John Barnes named it, was in use before the agreement of 29 March, perhaps as early as February, then both Hopwood's and Paul's statements are compatible with Paul being the main creator of the camera. On the other hand, the 1925 letter to Ramsaye would suggest that Paul was less responsible for the camera than he maintained—even to the extent of possibly having had it made by another specialist, as Richard Brown has suggested.[55]

But do we need to make a choice between these rival accounts—to become Acres or Paul supporters, as if they occupied opposing poles, like, say, Vertov or Eisenstein in early Soviet cinema? Short may have come to regret acting as go-between, but he continued working with Paul, which surely led to Acres's caustic reference to him in 1912. Acres would in fact achieve various undisputed "firsts" in his own right. He developed a projector before Paul succeeded with his, showing his Kinetic Lantern or Kineopticon privately in New Barnet in August 1895 and at the Royal Photographic Society on 14 January 1896. He also presented an all-film program publicly in a "picture palace" on Piccadilly Circus in March 1996, before this burned down "through the carelessness of an assistant . . . in his absence."[56] And in June of that year, he filmed the Prince and Princess of Wales in Cardiff, which led to an invitation to Marlborough House in London in July to show his films before a royal audience, gathered for the wedding of Princess Maud and Prince Charles of Denmark. His descendants and present-day supporters clearly feel that this is a proud record, and that his very lack of—or lack of desire for—commercial success with moving pictures vindicates him as a "true" pioneer, untainted by business compromises.

We are faced here with a choice of perspectives rather than a simple either-or judgment. From a developmental perspective, it seems clear that Paul would not have had the use of, or the results from, a film camera without the contribution of Acres—or perhaps some other experienced photographer. Like Edison, although on a much smaller scale, Paul had a workshop and a company behind him, and could draw on varied contributions to arrive at a workable apparatus. He was in a position to invest and innovate—and prepared to work exceptionally hard during the spring and summer of 1896, as we shall see—with the result that he prospered in the new industry. Acres too continued to innovate, especially with his Birtac portable camera of 1898, which was the forerunner of later narrow-gauge systems such as 9.5mm and 16mm, but did not prosper. From the evidence of Acres's writings, both published and private, between 1896 and 1912 he was at pains to discredit Paul and to claim that he had previously at least "designed" an early version of the apparatus that produced the well-known pictures of 1895. "Designed," perhaps, although not necessarily built. In Deac Rossell's opinion, Acres undoubtedly possessed "a vision of moving pictures" but "had no business sense."[57]

Acres refers in his 1912 notes to "documentary evidence in his posses-sion now (1912.Jan)" that the intermittent mechanism shown in Talbot's book "was not invented by Paul." He also mentions a "letter from Short," which appears to have been further evidence of Paul's perfidy. However, neither document seems to have survived. And even if they had, the issue remains one of perspective, since Paul did not claim priority but rather to have solved a problem by introducing a well-known mechanical device, the star-wheel or "Maltese Cross," to produce intermittent movement.[58] Acres seems wedded to priority—who came first?—within what we might call the "lone inventor" model. Paul, by contrast, takes a business and technical approach, underpinned by legality. Replicating the Kineto-scope was not problematic because the device was not legally protected. That Edison had tried to close down his Kinetoscope business did not prevent Paul making a business proposal to Edison in March 1895, to pool resources—any more than it prevented Edison from copying Paul-Acres films for sale in the United States. Contracting with a supplier of services (Acres) confers rights and penalties: if Paul is paying for Acres's services, he is entitled to full use of them; and if Acres breaks the agreement, by claiming sole patent on the camera developed while under contract to Paul, as he did on 27 May 1895, then Paul would be entitled to compensa-tion, or to take over their joint assets. In fact, neither party seem to have made any public reference to the incident that ended their agreement, and both apparently continued to use and improve the camera that had resulted, while producing their own projectors, suggesting at least a de facto recognition of a mutual right to continue using what had been cre-ated, albeit not together.

What light does the dispute shed on their characters? In some ways, they seem rather similar personalities, if differently focused and start-ing with different opportunities. Acres, born in the United States to En-glish parents in 1854, was fifteen years older than Paul and had lived an adventurous life.[59] After the early death of his parents, he was brought up by an aunt, who sent him to study in Paris, where he became inter-ested in art and photography. But he would return to America and live the life of a frontiersman and prospector for some years before head-ing back to Europe and England sometime in the late 1880s. This trans-atlantic pattern, if not the details, inevitably recalls the life of Eadweard Muybridge (1830–1904), who left England for America and became first a landscape photographer and then a pioneer of sequential or chrono-photography—and had an important influence on Edison's decision to explore moving-picture recording.[60] Acres seems to have developed his photographic expertise after reaching England, which led to a job with

Elliott and Sons of Barnet, well known makers of photographic plates, like the Lumière company of Lyon. Acres married Annie Cash in 1890 and apparently started working for the Elliotts in 1892, although whether he was manager of their works (according to his grandson Alan Acres) or a traveling salesman (Paul) is unclear. Perhaps he had become a salesman by late 1894, in the hope of earning more from commission. The pattern of his subsequent career suggests a restless search for both recognition—he was proud of being a fellow of the Royal Photographic Society—and business success.

Paul showed some of the same tendencies, but was more cautious and more successful—or luckier—in business. He was also active in professional and trade bodies, and eventually in historical scholarship. And as later events in 1895 would show, Paul had an imaginative or speculative ambition that compares with Acres's relatively conventional interest in nature photography. From the slender evidence available, both seem to have been demanding and single-minded, exacting toward their employees.[61] What may have seemed to Henry Short an ideal combination of mechanical and photographic expertise resulted instead in a clash of individualists, with very different conceptions of what was at stake. Yet they succeeded in their main aim—to produce a practical camera and usable films—which body of work would become the cornerstone of the moving-picture business in Britain, and further afield.

The First Films

The sixteen or so known films produced with the Paul-Acres camera in early 1895—leaving aside the disputed questions of agency and authorship—are of four main types: "performances," similar in principle to Edison's Kinetoscope subjects (performing bears, boxing kangaroo, skipping dogs, a boxing match, a comic shoeblack, and two sketches by a lightning cartoonist); national events, such as the Oxford and Cambridge boat race and the Derby; genre scenes (*The Carpenter's Shop*, *Rough Sea at Dover*); and finally, an intriguing proto-drama. There is also an experimental study or demonstration now known as *Incident outside Clovelly Cottage*, in which Acres's wife and their baby son's pram appear, as well as Henry Short wearing cricket whites—for maximum contrast against the dark brick background, it has been suggested. This effectively belongs to the same category as the *Experimental Sound Film* made by Dickson for Edison in 1894, to test or demonstrate linkage between the Kinetograph and the Phonograph. Neither seem to have been intended for pub-

4. *Rough Sea at Dover*, filmed by Acres early in 1895 for the Kinetoscope, and projected by Edison in his Broadway debut, April 1896.

lic showing, yet their very informality seems to hint at some hidden narrative, and to invite speculation.[62]

Clovelly Cottage has another claim to fame, since it was almost certainly the film Paul and his associates first succeeded in running through a Kinetoscope, late one night in Hatton Garden—an event that gave rise to the most famous anecdote of early British filmmaking. According to the first published account of the incident, mentioned by Paul in an after-dinner speech:

> It was 3am in the morning when [we] successfully finished [our] film and [we] made such a cheering over the fact that the police came in to know what was the matter.[63]

That this did happen was confirmed by Acres, in a marginal note on his copy of Talbot's *Moving Pictures*, where it was first retold, inaccurately, implying that the police had witnessed a *projected* picture. As Acres wrote, "The scene here described took place when the first of Acres' pictures were put in the Kinetoscope."[64] Did this mean he was present, witnessing it for himself, or merely heard about it from Paul while they were

5. Henry Short, in cricket whites, filmed as a test in early 1895 outside Acres's house in Barnet (later known as *Incident outside Clovelly Cottage*).

still collaborating? Whichever was the case, the story would take on a life of its own. Soon attributed to William Friese-Greene, it was eventually brought to life on screen as one of the most memorable scenes in *The Magic Box*, with Laurence Olivier playing the policemen to Robert Donat's Friese-Greene.[65]

Acres was almost certainly the cameraman for all of the Paul-Acres Kinetoscope films, and he continued to be proud of the fact that most were filmed in natural settings, even if the "performance" subjects were clearly arranged and staged for the camera.[66] Edison's Black Maria subjects had also been lit by sunlight, but the scenes were perceived as taking place on a kind of stage. All combined lively movement—the essential element in all films of the "demonstration period"—with traditions of performance that reached back through vaudeville and music hall to the early history of itinerant entertainers, as if this new technology were being grounded in a much older popular culture.[67] Topical events, such as the Derby and the Henley Regatta were also distinctive demonstrations of Acres and Paul's "animated photography" being able to do something conspicuously different from Edison's: instead of requiring that subjects be brought to the camera, the camera could "attend" great events on behalf of distant spectators.

But it was not only such obvious achievements that would impress the first spectators. As the Lumières would also discover, viewers were entranced by *evocative* aspects of the moving image—the movement of sea, wind, or steam. *Rough Sea at Dover* would remain one of the most popular subjects in Paul's and others' catalogues for years,[68] along with *Sea Cave near Lisbon*, which Short filmed for him in 1896. While "rough sea" was to some extent an established genre scene, popular with painters and printmakers—for audiences already attuned to maritime subjects, and no doubt for those who had never seen the sea, there was, it seems, an inexhaustible pleasure to be had from seeing those waves break repeatedly against Admiralty pier—as if the new form of picturing had come to pay its respects to the elemental scene of sea meeting land.[69]

The two films of the lightning cartoonist, Tom Merry, performing his popular stage act emphasize the links between the Kinetoscope and the music hall or fairground, where moving pictures would serve their apprenticeship, helping these institutions remain "modern," until the rise of extended narrative came to promote new specialist picture theaters. But probably the most surprising item in the original Paul-Acres repertoire is the drama *Arrest of a Pickpocket*.[70] Neither Acres nor Paul appear to have commented on this, and Acres seems to have favored documentary uses of cinematography. Yet he would insist in his 1912 notes that "the

6. *Arrest of a Pickpocket,* made for the Kinetoscope during Paul and Acres's brief collaboration in early 1895. National Fairground and Circus Archive, University of Sheffield.

first Cinematograph stage was erected by Birt Acres at 45 Salisbury Road, Barnet, England," which must have been for this pioneering crime drama, filmed against a sunlit backdrop. What is striking is the sure control of the three figures involved, amounting to highly effective fight choreography. The fact that Paul produced *Arrest of a Bookmaker* in the following year might indicate that he was also the moving force behind the earlier *Arrest,* so reopening the question of who was responsible for these 1895 productions.[71] But since Paul's later *Arrest* is lost, there seems to be no way of settling this, beyond assuming greater collaboration than partisan views have allowed.

Exhibition and the Empire

The business of showing films to the paying public is still known by its traditional name of "exhibition," which until the era of multiplatform exploitation (another traditional term) formed the last link in the chain that begins with production and reaches the exhibitors through distribution. None of these terms or businesses existed in the summer of 1895, yet they soon began to take shape. On 15 May, Paul mounted a "premiere" of "the first English scenes by the European manufacturer Mr Paul," in a tavern on Leather Lane, close to his Hatton Garden workshop.[72] But his main venture was to install, from 27 May, "fifteen [Kinetoscopes] at the Exhibition at Earl's Court, London, showing some of our first British films, including the Boat Race and Derby of 1895." Simultaneously, he took a step toward becoming a film distributor by advertising for sale "New Topical Subjects" in the *English Mechanic.*[73]

We can of course speculate, in the absence of any further detailed information. Did Paul find he had a surplus of Kinetoscopes, having tooled up for mass production?[74] Or had he come to realize that the takings from exhibiting the machines, even if shared with "business friends," were likely to be much greater than the profit from merely selling them—especially since he was selling at less than Edison's price?[75] Fifteen Kinetoscopes, charging 2 pence per view, with each of these lasting less than a minute, meant a potential return of £7 per hour or up to £50 per day—equivalent to an astonishing £7,000 in modern terms. Paul recalled "the sight of queues of people, waiting their turn to view," so we can assume that this was not mere speculation. And thanks to the partnership with Acres, he had distinctive British subjects to publicize, including two of the major annual sporting events that still define the structure of Britain's national life.

The venue for this first venture into film exhibition was the *Empire of India* exhibition at Earls Court showground in West London, which ran from May to October 1895. This was the creation of the Hungarian-born impresario Imre Kiralfy, as was the new Earls Court site itself, which Kiralfy had rebuilt in the style of Chicago's White City, laid out for the World's Columbian Exposition in 1893. Appropriately, Kiralfy had first made his name with an elaborate staging of Jules Verne's *Around the World in Eighty Days* in the United States in the 1880s,[76] before moving on to re-create the ancient world in such shows as *The Fall of Babylon* and *Nero, or The Fall of Rome*. In Chicago, he had created nothing less than *America*, staged at the Auditorium Theatre, and his first major production in London was a spectacular re-creation of Venice, combining theater and exhibition amid a network of canals, at Olympia in 1891. These successes led to his investment in building a new exhibition ground at Earls Court that would offer even greater scope for spectacle.

Empire of India launched the theme that would dominate Kiralfy's last decade—the empire of his final home. Like his fellow countryman three decades later, the film producer Alexander Korda, Kiralfy seems to have had an even greater enthusiasm for empire than the native British. Both of these showmen would give British audiences an unabashed, though shrewdly conceived, celebration of empire, stretching from its zenith in the mid-1890s to its twilight in the 1930s.[77] Paul would also make a major contribution to celebrating the British Empire with his films of Queen Victoria's Diamond Jubilee and of the Delhi Durbar, but in 1895 his more modest role in Kiralfy's exhibition-spectacle was to inject a note of modernity. Electricity played an important part in these spectacles, illuminating the whole exhibition area and powering the Great Wheel

7. Guide to Imre Kiralfy's *Empire of India and Ceylon* exhibition at Earls Court. Paul
had exhibited Kinetoscopes during the spectacle's first season in 1895, before Ceylon
was added to the title, and projected hand-colored films during the summer of 1896.

that towered over Earls Court for a decade after the show. So while India was visualized in the paintings of the American oriental specialist Edwin Lord Weeks and in mocked-up Mogul architecture, the Kinetoscope demonstrated the electrified future of spectacle. Ironically, what Edison had failed to have ready for the Chicago Columbian Exposition was now present at London's equivalent, albeit in a locally produced form.

The exhibition's success led Kiralfy to extend its run, "adding Ceylon and other Crown Dependencies in Asia to this already instructive and interesting Exhibition."[78] Paul, however, had no reason for continued involvement, having embarked on a new and hectic career as a projected picture exhibitor in early 1896. The *Empire of India* exhibition, he would later claim, "first caused me to consider the possibility of throwing the pictures on the screen."[79] And by the autumn of 1895, another contemporary stimulus also fired his imagination, pointing the apparently sober young engineer toward an even more ambitious goal, which seems today like a premonition of his impending career in filmmaking, and indeed of the promise of cinema itself—a "time machine."[80]

✳ CHAPTER 2 ✳

Flashback

AN ENGINEER'S EDUCATION

If Paul was briefly beguiled by the idea of a "time machine," he was guided by logic in discerning the limitations of the Kinetoscope's single-viewer model and seeking to move beyond it, and was soon at work on creating a lantern-based projection system. In fact, he was the only one among the pioneers—Edison, Acres, Lumière—who would directly engage in "theatrical" (as it is still called in the film industry) exhibition, distribution, *and* production, all within a matter of weeks in mid-1896.[1] Yet he has remained a surprisingly shadowy, little-known figure. In one of his letters to Oskar Messter in 1935, he admitted to being "rather averse to personal publicity." So averse, that he left no personal or company papers, and his family background has proved barely traceable. But we can place him professionally near the center of a technological revolution that began before moving pictures with the new science of electricity. And this is what gave him the skills, and no doubt the confidence and early financial security, to pursue the new opportunities the Kinetoscope offered.

An Elusive Family

What little is currently known about Robert William Paul's personal life comes from two main sources, other than the obituaries that followed his death in 1943. One is the routine documentation that records his birth, marriage, and death, supplemented by two census returns from 1901 and 1911, some street directory entries, and a bare handful of letters. The other, more tantalizingly, is an unpublished manuscript by a friend of Paul and his wife that seems to have been begun in the 1950s and reworked until as late as 1996.[2] Irene Codd claimed that "Mr and Mrs Paul were friends of my parents and he was a scientific colleague of my late father," and although her text incorporates unacknowledged material from other sources, it includes some anecdotes and observations that are undoubt-

8. Albion Road, Islington, where Paul was born in 1869. These houses, opposite the site of Paul's birthplace, retain the original character of this North London street, near Highbury Corner. Photo: Ian Christie.

edly based on firsthand acquaintance with the Pauls.[3] In the absence of other sources, these provide a valuable extension of the bare record of Ellen's and Robert's lives together.

Robert William Paul was born on 3 October 1869 at 3 Albion Road in North London, the first child of George Butler Paul and Elizabeth Jane Lyon.[4] The birth certificate records George Paul's occupation as "merchant clerk." He would later be described as a shipowner, but there do not appear to be any records of ships being owned until 1898. Six small merchant ships were registered as belonging to "Paul and Shellshear" between 1898 and 1907, but even if he was not an owner, he was certainly involved in the shipping business during the intervening years. One obituary of Robert referred to him "travelling the world" on his father's ships as a boy or young man, hinting at what his friends may have known that is now otherwise unverifiable. However, the fact that he later traveled to Sweden and more than once to Norway fits well with George Paul's coming from Lincolnshire, near the port of Hull, and his known ships apparently being involved in the Baltic trade.[5]

Elizabeth Lyon was born in France, to British parents, in 1842. Her father appears to have been a clergyman and the family to have originated from Newcastle-upon-Tyne. Soon after Robert's birth, in a street near the busy junction of Highbury Corner in Islington, the family was on the move. Their next child, Arthur, was born at Stratford in East London in 1873, and by 1901 was a ships' stores dealer, apparently having entered his father's business. A third son, George, followed in 1877 and is listed as a shipowner's clerk in the 1901 census. Between these sons came the family's first daughter, Elizabeth, whose birth took place in Newcastle in 1875, suggesting that her mother might have been staying with her own family.

The Paul family was living in Stratford, on the eastern margin of London, for at least part of the 1870s, in an area that had been agricultural before Victorian planning legislation banned "dangerous and noxious industries" from the city, and made it a center of industry and engineering. In 1886 the *Times* referred to it as "London over the border" and described the area's new character:

> Factory after factory was erected on the marshy wastes of Stratford and Plaistow, and it only required the construction at Canning Town of the Victoria and Albert Docks to make the once desolate parish of West Ham a manufacturing and commercial centre of the first importance and to bring upon it a teeming and an industrious population.[6]

The Pauls' last child, Alice, was born back in London in 1880, in "Stampford Hill," presumably what is known today as Stamford Hill.[7] By 1901 Robert's parents were living in Loughton, on the edge of Epping Forest, and it was here that his mother died in 1906. Did this location mean they were prospering? Loughton had a dual reputation: as a genteel area from where city folk could easily reach their businesses in London and, after the coming of the railways, as a popular resort for Cockneys in search of a cheap day out—known as "Lousy Loughton" due to the lice and fleas these trippers brought. From 1881 the Ragged School organized trips for deprived London children to Loughton. It was also known as a staunch nonconformist area, although there is no evidence of the family having

9. Paul entered the City of London School in 1883, the year that its new building opened on Victoria Embankment.

had any religious affiliation. Despite this, there were recurrent suggestions that Paul was Jewish, apparently based only on his appearance. An apprentice recalled him being "of obvious Jewish extraction."[8]

Both Robert and his father prospered during the 1890s, while the younger sons seem to have fared rather less well, apparently not having benefited from Robert's level of education. Both Arthur and George served with the elite City Imperial Volunteer regiment after the Anglo-Boer War broke out in 1899. They survived, and were presumably back in England in time for the 1901 census. Did either of them serve again, during the Great War? Did their health suffer from being on the Cape, or later on the Western Front, or from the great influenza epidemic of 1918–1919, which killed as many as the war had? There is no way of knowing. George died in 1919 in Bournemouth, aged forty-two, recorded as a retired bank clerk of "no fixed residence." His brother Arthur was "in attendance," but he would die only three years later, aged fifty, in Coventry, leaving a widow. Arthur is described as a typewriter salesman on his death certificate, so neither of the younger brothers had stayed in the shipping business. Their father would outlive both, not dying until 1930, and leaving a substantial fortune. The last we hear of the daughters is a provision for some legacies in Robert Paul's will.

As the eldest son, Robert seems to have benefited from a better education than his parents or his siblings. He entered the City of London

School at the age of sixteen, in 1883, when this unusual public school—which admitted nonconformists and Jews—had recently moved from Cheapside into a new building on Victoria Embankment.[9] Here he had a rare opportunity to study science at the first secondary school in Britain to introduce this subject, alongside the traditional academic subjects. It can hardly be accidental that one of his first films, *Blackfriars Bridge* (1896), is taken from an angle that shows the imposing City of London School building in the background. From the City of London, he would progress to the newly established Finsbury Technical College in 1885, where in his second year he was "elected Senior Student after passing the College course in [the] Electrical Engineering Department" and won a prize.[10] The same note records that he had "accepted the position of works manager to the firm of Macdonald and Co, manufacturing electricians of Bow." In the following two years, Paul would pass rapidly through a number of companies. According to Irene Codd, this was in spite of his father's wish that he should follow him into the shipping business, as his brothers would do, at least temporarily. Despite paternal opposition, Paul had clearly glimpsed an exciting new prospect in electrical engineering—little realizing that it would lead to a pioneering role in an even newer industry.

Electrical Invention

Electricity had, by the mid-1880s, become the most exciting technology of the late Victorian world. Building on the foundational research into electromagnetism of Humphrey Davy and Michael Faraday at the Royal Institution, carried on by William Thomson, later to become Lord Kelvin, inventors had begun to harness the power of electricity. Telegraphy was the initial main field of application, and signals were sent through the first successful transatlantic submarine cable in 1858, using a mirror galvanometer invented by Thompson that could record minute variations in current. This cable would prove short-lived, and regular transatlantic telegraphy was not reestablished until 1866, when Thompson produced a further device, the siphon recorder, later followed by the automatic curb sender, which became the standard means of sending submarine cable messages.[11]

Faraday's work on the measurement of electrical current had led to a series of instruments, initially intended for experimentation but soon proving to have practical and commercial potential. Little wonder that Paul, as part of the first generation of trained electrical engineers, would

revere Faraday and make a major contribution to the 1931 centenary celebration of his discovery of electromagnetic induction.[12] In the 1860s and
1870s, what would later be called "technology" was still intertwined with
"natural philosophy" and science, with Thomson being drawn into engineering by the onward rush of electrical innovation. Ever more sensitive
and specialized forms of measurement were needed, both for the laboratory and for emerging electrical industries; hence the new profession of
electrical instrument making.

The other great application of electricity, which had perhaps more immediate public impact than telegraphy, was in lighting. There had been
experiments with incandescent electric lighting in the 1840s, but in 1878
Edison made it his main priority after the Phonograph, and the following
year succeeded in encasing a carbon filament within an evacuated glass
container—the prototype of the modern lightbulb. As with most "inventions," there were competing claims to priority—in the case of the lightbulb, J. H. Swan and C. H. Stearn in England—but increasingly Edison's
name would prevail in the popular mind. This was partly due to his assiduous and inspired cultivation of the press, but also to his various companies making use of the many exhibitions in Europe and the United States
that thrived on displaying innovation. Edison's electric lighting and distribution systems were on show at the International Electrical Exposition
in Paris in the summer of 1881, where "the completeness of its conception
made a profound impression on the foremost European electrical engineers of that era."[13] In the following year, London's International Electric
and Gas Exhibition—housed in the re-erected Great Exhibition building
known as Crystal Palace—featured an even greater range of Edison inventions, including

the steam dynamo; specimens of street pipes and service boxes used in
the Edison underground system of conductors, and the system of house
conductors with devices for preventing abnormal increase of energy in
house circuits; apparatus for measuring the resistance of his lamps, for
measuring the energy consumed in lamps, and rheostats for restoring currents; also thermogalvanometers, carbon rheostats, dynamometers, pho-
tometers, carbon regulators, Weber meters, current regulators, and circuit
breakers for controlling electric light circuits … the carbon telephone, the
musical telephonograph, telephone repeater, and numerous apparatus for
demonstrating the method of varying the resistance of a closed circuit by
contact with carbon, illustrative of the experimental factors of the Edison
carbon transmitter. Incandescent lamps, the process of the manufacture
of lamps, and various designs of electric light chandeliers[14]

Edison had created the first public electricity supply station in New York in 1881, and a similar station opened in London in 1882, located in the basement of a house in Holborn Viaduct and powered by the "Jumbo" turbine. Companies were being created to bring electric lighting to individual districts, while Edison's staff and agents were seeking opportunities for public display.

It was a time when "all electrical invention seemed to be made," as another veteran of this period later recalled.[15] Hugo Hirst had come to England from Germany at the end of the 1870s, and began working for the newly created Electrical Power Storage Company in London in 1881. He remembered the general public ignorance as to what electricity was, and such stunts as fitting up the Gaiety Theatre "for no payment but simply to have the advertisement for what the accumulators could do." He also told of starting to work for the Manchester Gas Lighting Company in London, when the Electrical Power Storage Company "inevitably" went into liquidation, and of being advised not to go to Australia on the company's behalf in 1883 because "there will be more electricity within four miles of Charing Cross than in the whole of Australia."[16] There would indeed be more electricity, and much need for a new generation of skilled electrical workers—although this was only one factor contributing to a widely perceived crisis in British science and industry.

The country that had engaged with modern technologies earlier than any other, adopting innovations in mining, metals, textiles, and transport during the later eighteenth century, began to realize that it had fallen behind other nations precisely in these fields where it had once led. Charles Babbage's *Reflections on the Decline of Science in England*, based on his difficulty finding support to develop the earliest form of computer, the "difference engine" in 1830, helped to launch a national debate. And the same theme returned after the Great Exhibition, when Lyon Playfair and T. H. Huxley urged that technological development merited not only a permanent display but the establishment of a college devoted to technical education on its site.[17] Sixteen years later, the debate had a new urgency, when British's performance at the 1867 Paris exposition was judged a national disgrace after, according to John Scott Russell, "we were beaten, not on some points, but by some nation or another at nearly all those points on which we had prided ourselves."[18] A Royal Commission investigated scientific education in schools in 1871, and congratulated Rugby on being the first among Britain's elite public schools to offer science classes, while concluding that technical education was better managed in many other countries.[19] But perhaps the decisive impetus came from a speech by William Gladstone in 1875, when he called on the traditional Lon-

10. Finsbury Technical College, the first of its kind in Britain, opened its Leonard Street premises in 1893, where Paul would demonstrate the Theatrograph three years later.

don craft guilds to regain their original purpose of developing the "crafts, trades and 'mysteries,'" rather than survive as mere ceremonial bodies.[20] With the prospect of further inquiry into their activities and considerable wealth, the guilds came together to form the City and Guilds of London Institute for the Advancement of Technical Education in 1878. This led to the creation of two major new institutions: Finsbury Technical College, located near the City, and the Central Institution, established in South Kensington on the former Great Exhibition site, which would eventually become, after several amalgamations, the modern Imperial College of Science and Technology.[21]

The first director of the new institute, Philip Magnus, was a part-time rabbi when he began the task of forming the new colleges, but he had also studied the German vocational education system, which was widely considered superior to any provision in Britain. Among the first professors he secured for Finsbury were Henry Armstrong, a former student at the (privately run) Royal College of Chemistry, and William Ayrton, an electrical engineer who would become Robert Paul's first mentor and patron. In the absence of any insight into the young Robert Paul's motivation, it may be useful to consider the better-documented route taken by one of his near-contemporaries at the college, who would also become a notable instrument maker.

William Taylor was born in 1865 in Hackney, to parents of modest though respectable status, and from early childhood showed great mechanical ability.[22] Learning from the village blacksmith and local wheelwright, and from articles in the *Edinburgh Encyclopedia* by Sir David Brewster, inventor of the kaleidoscope and a version of the stereoscope, William and his brother set up their own workshop, where they improvised "a pair of the first telephones ever made in England and one of the first copies of Edison's Tinfoil Phonograph" (a reminder that copying one of Edison's "inventions" had launched other careers). The boys were lucky to attend Cowper Street School, Finsbury, where Dr. Richard Wormald was already implementing the Devonshire Commission's call for more science education, and William acted as a demonstrator for his teacher before going on to become one of the first students at Finsbury Technical College in 1879. At this point Taylor's path becomes strikingly similar to Paul's. After leaving the college, he was apprenticed to Paterson and Cooper, where he worked on the fixed-coil ammeter designed by his former professors Ayrton and John Perry in 1883 — anticipating the Ayrton-Mather moving-coil galvanometer that Paul would manufacture from 1891.

The company in Bow that Paul joined as works manager must have been Coates and Macdonald, from which Hugo Hirst and his new General Electric Apparatus Company bought their switches.[23] But he cannot have stayed long, since we know he also worked at two other important companies in the electrical field: Elliott Brothers and the Bell Telephone Manufacturing Company in Antwerp, Belgium. Elliotts had been in Central London, in St. Martin's Lane, for nearly a hundred years, making mathematical and traditional scientific instruments, before moving into electrical instruments for telegraphy and the other new applications. No doubt the company would have wanted Paul's up-to-date expertise, and perhaps his continuing contacts with Ayrton, Perry, and Thomas Mather. Interestingly, Elliott Brothers would move from St. Martin's Lane to what was then a greenfield site in Lewisham, Kent, building its large new Century works the same year Paul acquired land in North London for the expansion of his filmmaking business.[24]

We have no detail or chronology for Paul's employment with Coates and Macdonald, Elliott Brothers, or the recently established Belgian branch of America's Bell Telephone Company.[25] Since he established his own company in 1891, he cannot have spent more than about thirty months at these companies, which suggests that his aim was to gain experience rather than make a career with any of them. Irene Codd provides some interesting and plausible personal background to this period. Apparently the twenty-one-year-old Paul suffered from looking too young for the level of responsibility he had acquired, which led him to grow the "little square beard that he wore all his life." More importantly, she reveals that Paul had a major falling out with his father, who again asked him to join the shipping business and refused to back his son's proposed electrical instrument company. According to Codd, Robert met another shipowner, a Mr. Knight, who offered to support the venture by buying him secondhand machine tools and helping to find premises. Their search led to the Robert W. Paul Instrument Company being launched on the second floor of 44 Hatton Garden.

A New Industry

Paul's fledgling company must have depended heavily on his prior contacts in a still-new arena, and Codd describes him initially doing the rounds in search of repair work. Nor was he alone, as Hirst's reminiscences of the early electrical business make clear. There were others willing to take a risk in this exciting new field, which was clearly on the brink of huge expansion as electricity started to become a part of everyday life in Britain. The new industry needed instruments, to measure and control power supply and to regulate many kinds of installation, from public lighting to workshops, factories, and transport systems.

A struggle was still raging between supporters of direct and alternating current, conveniently known as DC and AC. This battle can be, and often has been, dramatized as between Edison and the Nikola Tesla — with Edison painted as a ruthless promoter of DC, even to the point of damning AC by simulating the electric chair in an early film and actually filming the electrocution of an elephant.[26] Tesla, by contrast, is often cast as a true scientist and visionary, hampered and persecuted by Edison — and portrayed as a kind of magician, spectacularly by David Bowie in Christopher Nolan's 2006 film *The Prestige*.[27]

The reality, of course, was more complex, with scientific and industrial considerations ranged on both sides of what was indisputably a contest. While alternating current, which periodically changes polarity between positive and negative, is highly suited to widespread distribution and eventually became the basis of most supply systems, its higher voltage makes it theoretically more dangerous and less suited to many applications requiring a steady direct current. In practice, the same insulation precautions are needed for any high-voltage systems; AC, however, has an advantage in that it can be "transformed" down to the power levels needed for, especially, domestic purposes. Edison's campaign for DC was in defense of his patents and the large revenue they were starting to return; but after Tesla's series of demonstrations and patent successes in the late 1880s, the tide began to turn. AC was widely adopted after the industrialist George Westinghouse acquired Tesla's patents, and its victory was confirmed by the decision to use it for the Niagara Falls generating plant in 1893.

This wider background should help us keep the early business of moving pictures in proportion. Buildings and whole districts being lit by electricity was undoubtedly a larger drama in most people's lives that the arrival of "living pictures." And yet, through the name of Edison — already synonymous with technological wizardry and with the application of electricity — moving pictures came to be associated with the electrical revolution. Edison's Kinetoscope was battery-operated, which fact made it problematic, according to Paul, "due to the difficulty, at that time, of recharging its accumulators."[28] Ironically, the next development in moving pictures would *not* be electrical, since Paul's, Acres's, and the Lumières' projectors were all at first hand-cranked. Nor did they require electric illumination, since gas was usually preferred in the 1890s and early 1900s, although the electric arc would later be used in larger halls that needed a more intense light source.

Mentors and Patents

Several of Paul's teachers at Finsbury Technical College were leading players in the "current wars" of the 1890s and were also involved in the many legal battles over patents that accompanied the electrical revolution. At a time when academic posts in science were still rare, their

11. William Ayrton, first professor of physics and telegraphy at Finsbury Technical College, had introduced electrical studies to Japan in the 1870s. Paul would produce instruments designed by him in the 1890s.

experience before coming to the City and Guilds' colleges was varied. One of the college's founding professors, William Ayrton (1847–1908), had studied under Thomson in Glasgow before going to work for the Indian telegraph department in 1868 and spending five years in Bengal. After a year with the Great Western Railway, he became the first professor of natural philosophy and telegraphy—an oddly compound title that reflects the transition in science then underway—at the Imperial Engineering College in Tokyo. While there, he introduced electric arc lighting to Japan and helped educate the generation that would launch many of Japan's new electrically based industries, including Ichisuke Fujioka and

Shoichi Miyoshi, the founders of the Tokyo Electric Company, later to form part of Toshiba. He also formed a lasting professional partnership with another guest professor, the engineer John Perry, which would continue when they both arrived at Finsbury Technical College in 1881.

Like Thomson, Ayrton and Perry did not hesitate to involve themselves in issues of the day, such as high-voltage power transmission, and practical applications, including railway electrification and the electric tricycle. Ayrton also continued Thomson's concern with measurement and devised a series of instruments, which ex-students of the college, such as Taylor and Paul, would bring to market. The first of these was a series of ammeters, to measure the flow of current, which began with a permanent-magnet model in 1880, followed by the magnifying-spring model in 1883, both designed with Perry. They would also design a dynamometer and a wattmeter, and in 1891, Ayrton and another Finsbury colleague, Thomas Mather, designed a galvanometer, which measured current on a scale—a requirement brought about by the increasing use of large currents in lighting systems.[29] It was this that Robert Paul's new company produced and began to market, using modern manufacturing methods.[30] Later, Paul would contribute to this tradition by creating what has been described as his own "masterpiece" of instrument design, the unipivot galvanometer, but in 1891 the Ayrton-Mather galvanometer must have helped draw attention to the new firm, emphasizing Paul's continuing links with leading figures in the electrical world.

The other key figure at the college with whom Paul remained in contact was Sylvanus Thompson, who arrived as the new principal in 1885, the year Paul started. Thompson was a prolific and extrovert character, whose lasting legacy was as an inspiring teacher and popularizer of science. A Quaker and pacifist, his early education was in classics, but after attending a lecture by Sir William Crookes in 1876 he was inspired to take up science and became a lecturer in physics at the newly founded University College of Bristol, and eventually its first professor of physics.[31] However, his love of teaching, especially young people and lay audiences, and desire for a more practical environment, drew him to Finsbury Technical College, where he remained until his death in 1916. Once there, he threw himself into organizing an unusual, intensive curriculum for its two main types of student:

> Apprentices, journeymen and foremen who desire to receive supplementary instruction in the art and practice, and in the theory and principles of science, connected with the industry with which they are engaged.

Pupils from middle-class and other schools who are preparing for the higher scientific and technical courses of instruction to be pursued at the Central Institution.[32]

We do not know whether Paul entered as an "apprentice" or a "pupil from a middle-class school," or as one of Finsbury's anomalous students—who would include Bertha Marks, a Cambridge graduate frustrated by limitations placed on women in the traditional universities. Marks studied with Ayrton before becoming his second wife in 1885 and going on to a distinguished scientific career in her own right.

It was indeed a time of opportunity, with new possibilities for class mobility and enterprise, but as Hugo Hirst recalled, explaining how he kept his social and business life separate, "It was simply awful in those days to be a shopkeeper."[33] The Finsbury College Old Students Association would doubtless have stood Paul in good social stead, and in 1890 he became an associate member of the newly formed Institute of Electrical Engineers, of which Sylvanus Thompson became president in 1899.[34]

Thompson was a vociferous controversialist, unafraid to enter any scientific domain. He was quick to take up new discoveries, such as X-rays and radioactivity, and was active as an inventor, with contributions to the telephone and a system of electric traction for tramcars. However, his improved telephone was eventually challenged by Edison's lawyers and his patent disallowed on a disputed term.[35] Ayrton would have similar experiences, appearing as an expert witness in many suits and countersuits relating to electrical patents. Such activities by his recent teachers would certainly have been familiar to Paul, and must have given him a working knowledge of the complex and often bitterly contested field of patents in new technology, even before his involvement with the Kinetoscope and with Acres.[36]

Thompson's wide interests and passion for popularizing science led him to give the traditional Royal Institution Christmas Lectures on two occasions, and for one at least Paul helped.[37] He recalled filming for Thompson

hundreds of diagrams, illustrating lines of force in changing magnetic fields [which] we converted into animated pictures by the one-turn one-picture camera which we had employed for animated cartoons.[38]

This is in fact the only indication we have that Paul's studio ever made animated cartoons. But as well as working with Thompson, he had filmed "pictures of sound wave 'shadows' and falling drops" for two well-known

12. Hatton Garden, where Paul started his electrical instrument business in 1891, photographed in the same decade. Now the center of London's diamond and jewelry business, the street then had a more varied character. Hiram Maxim, who invented the machine gun in 1881, was a near neighbor of Paul's. And the long-established instrument maker Short and Mason, specialists in barometers and compasses, was at number 40. Photo courtesy of London Metropolitan Archives.

scientists, C. V. Boys and A. M. Worthington (discussed further in chapter 10). The former remains famous for his classic study of the physics of soap bubbles, and the latter for an 1894 lecture on "The Splash of a Drop and Allied Phenomena," which was illustrated with some early instantaneous photographs.[39] Both continued to experiment and seek more instantaneous forms of photography to record ballistic and splash phenomena, and it is possible that Paul's film frames have circulated anonymously in the documentation of their work.[40]

Hatton Garden

Paul's new company was based at 44 Hatton Garden, a distinctive thoroughfare off Holborn, then with a dense mixture of small manufacturers and retailers, now all but monopolized by the diamond and jewelry business. Its police court had featured in Dickens's *Oliver Twist* (1836), and behind it lay the once-notorious Saffron Hill, where Fagin's "rookery" of thieves was situated. By 1890 Hatton Garden had become more respectable, housing a variety of businesses. Among its more colorful recent occupants was Jaques and Son, a long-established games company that had launched the modern Staunton chess figures and introduced both croquet and table tennis to Britain. Another was Hiram Maxim, who had developed his famous machine gun in Hatton Garden during the 1880s but was now involved with heavier-than-air flight. Maxim had been a self-taught instrument maker and electrician before making his fortune with the quick-firing gun. Elsewhere in Hatton Garden, Johnson and Mathey had become central to the fast-developing international gold business, recently boosted by major finds in South Africa and Australia.

Here Robert Paul plied his trade, producing instruments for teaching and research in electricity, as well as for the growing electrical trade. The record of his admission to the Institute of Electrical Engineering includes the surprising information that he exployed "about thirty hands," which suggests that the business was thriving.[41] We might wonder, however, about his status in a world that was changing but still governed by rigid class codes. Seventy years earlier, Faraday had made the painful transition from servant to gentleman, through exceptional talent that led to the creation of a special professorship for him at the Royal Institution. And it was this same uniquely English institution that had changed Sylvanus Thompson's vocation, and would later play an important part in Paul's life. The two City and Guilds' colleges, Finsbury and the Central Institution, set out to professionalize science education but could not themselves change the class structure that was acknowledged in Finsbury's admission categories.

The Theatrograph

Film historians have sometimes wondered why Paul took so long to develop a projector after producing the camera used for the filming earlier in 1895, apparently forgetting that moving pictures was never his only business. When he had started to make substantial profits from Kinetoscopes and their operation in 1895, he was forced to take on a second workshop for this business, on Saffron Hill. His speculative "Time Machine" patent in October of that year (of which more later) suggests that he was unsure how to develop this unexpected new business and was considering different possibilities. There was certainly no shortage of others aiming to capitalize on the appeal of moving pictures that the Kinetoscope had demonstrated. Acres wrote that "the Lumière apparatus was well known before Paul had attempted to do anything," and indeed the Cinématographe had been undergoing demonstration in France since March.[42] Acres himself was at work on his Kineopticon projector, patented as early as May 1895, and had given a "semi-public" demonstration in August, at the Assembly Rooms in New Barnet. Other inventors were active in Germany and the United States. But rather than pursue the chimera of "who was first," which has so dominated (and distorted) the early history of moving pictures, it is more relevant to point to the very obvious reasons why projection would attract so many pioneers, in view of the wide currency of the magic lantern, now increasingly known as "optical."

Devices for projecting images on a screen had been known since the seventeenth century, when early versions of the standard lantern were proposed by Athanasius Kircher and Christian Huygens.[43] By the mid-nineteenth century, photography and lithographic printing had made lantern slides cheaper and more plentiful, while new forms of illumination enabled lanterns to produce a brighter image on larger screens. A double or triple lantern using limelight could produce an image suitable for venues as large as Carnegie Hall in New York, Chicago's Symphony Hall, or the Free Trade Hall in Manchester, where leading lantern lecturers regularly appeared before capacity audiences.[44] Given the scale of the lantern industry, with its numerous equipment producers and slide suppliers serving the needs of thousands of presenters—many of whom would become the first users of film—the issue of film projection needs to be understood more as an offshoot of an existing practice than as an entirely new development. Certainly the film transport mechanism required innovation, but the other essential components of the system—illumination, lens, and screen—were already widely available.

We may never know whether Paul was spurred on by news of the Lumières' public demonstration in Paris on 28 December 1895, or by Acres's presentation of the "Kinetic Lantern" (a name that emphasized the apparatus's close links with the magic or optical lantern) to the Royal Photographic Society on 14 January 1896, when he showed some of the films made during his partnership with Paul, as well as new subjects taken in Germany.[45] In fact, Paul's immediate motivation may have derived from his existing business, as he explained to Oskar Messter in 1932:

Lack of capital prevented me from producing the [Wells-inspired "time machine"] pictures, but I proceeded to design the Projector. My desire

13. Ayrton-Mather galvanometer, manufactured and sold by Paul in 1895. Oxford Museum of History of Science.

was to produce an attachment for existing lanterns, chiefly to be used in Schools and Colleges, and to be sold for £5. This was shown at Finsbury Technical College ... and also at the Royal Institution.[46]

These demonstrations of Paul's projector, the Theatrograph, took place on 20 February and 28 February respectively, the former on the evening of the Lumière demonstration at the Polytechnic, and the latter in an "exhibition" in the Institution's library at which William Friese-Greene also gave a presentation.[47] The choice of venues was entirely consistent with Paul's stated intentions. While it may seem surprising that he had not grasped the entertainment potential of projected pictures after the *Empire of India* success, we should remember that his main business was now selling apparatus for educational and scientific use. As it happens, £5 was also the price of an Ayrton-Mather galvanometer, such as the one bought by a chemist, Herbert McLeod, in 1894.[48]

Moving pictures on the screen were nonetheless newsworthy, and Paul's demonstrations attracted attention, in spite of their less public venues, which could hardly have happened without him issuing advance notice. The Finsbury College showing was noted in the *English Mechanic* of 21 February, together with a notice of the Cinématographe being demonstrated on the same day at the Regent Street Polytechnic.[49] A report of the Finsbury show appeared in the *City Press* on the following day, and the Royal Institution demonstration was noted in several papers on 29 February. According to the *Daily Chronicle*:

> The audience sees upon a screen living pictures ... ships coming into the harbour, waves breaking on the shore ... all in a highly realistic manner.

Paul was evidently interviewed for this report, from which it appears that he still had the Wellsian "novel means of presenting" in view:

> He has some starting ideas in his head ... nothing less than, we understand, a vivid realisation of some of the imaginative scenery pictured in Mr Wells' *Time Machine*.

Questions inevitably arise about these events and reports, as they have become frequently cited landmarks. How far ahead did Paul know about the impending Lumière demonstration in London, and did he arrange his Finsbury demonstration to coincide as closely as possible? Or might friends at the college have invited him, since there was some rivalry between Finsbury College and Regent Street Polytechnic, which Sylvanus Thompson had described sarcastically in 1890 as an "excellent social club"?[50] Those suspicious of Paul's later recollections have also questioned his claim not to have realized the entertainment potential of moving pictures until after the Royal Institution show. What is not in doubt, however, is that at least one audience member at that second demonstration did grasp this potential and acted quickly to secure it.

Lady Florence Harris was the wife of the most important theater impresario in London, Sir Augustus Harris.[51] Famous for his extravagant Drury Lane pantomimes, Harris also had interests in other entertainment venues, and when Paul responded to a telegram inviting him to breakfast, he was invited to present the Theatrograph at Olympia, in an entertainment annex to an exhibition center, the Palmarium, then being managed by Harris. According to Terry Ramsaye, writing in 1926, and Leslie Wood, in 1947, Harris had already seen the Lumière Cinématographe in Paris and was keen to book a similar attraction while it was still a novelty.[52] These and other popular accounts stress how fleeting Harris and Paul thought the interest in projected pictures might be—possibly cases of "dramatic irony" dating from the height of the cinema's popularity. At any rate, a daily program of screenings began at Olympia on 21

IN THE PALMARIUM.

THE

THEATROGRAPHE

(PAUL'S)

ANIMATED PICTURES,

IN BRILLIANT COLOURS.

The most wonderful Scientific Marvel of the Age.

THE SAME AS NOW BEING SHOWN AT THE ALHAMBRA.

MARVELLOUS MOVING PICTURES FROM REAL LIFE.

ADMISSION SIXPENCE.

IN THE PALMARIUM.

THEATRE   ROYAL,   RICHARDSON,

"Hamlet in a Hurry,"

STYLLE RUNNYNGE

Hamlet   WHIMSICAL WALKER.

OSCAR   CARRE'S

Royal * Dutch * Band,

Conductor   -   Mr. C. Weichmann.

14. Advertisement for Paul's "Theatrographe" program in the Palmarium hall of Olympia, near Earls Court, June 1896.

March, with tickets priced at sixpence for five or six subjects, and Paul and Harris apparently splitting the take fifty-fifty.[53]

Two days earlier, screenings with a Paul projector also took place at the Egyptian Hall in Piccadilly.[54] This popular venue, opened in the early years of the century as a museum of curiosities, was now being run by the magician Nevil Maskelyne and billed as "England's home of mystery," with the young Georges Méliès a frequent visitor during the time he had spent in London in 1883. The appearance there of the Theatrograph, presented by the magician David Devant, forged an early relationship with magicians which would take it to a wide public, as the initially skeptical Maskelyne became a strong supporter, and probably a valued adviser, to Paul on the entertainment world.[55] Meanwhile, Acres also launched daily shows of his Kineopticon on 21 March, at Piccadilly Mansions; the first regular presentation entirely devoted to moving pictures, it ran until 10 June—a successor to De Loutherbourg's Eidophusikon presentation in nearby Lisle Street in the 1780s, and a forerunner of later "news cinema" programs.[56]

The adoption of the Theatrograph for entertainment purposes gath-

15. The Egyptian Hall in Piccadilly, shown here around 1900, opened as a museum of curiosities in 1812 and hosted many nineteenth-century novelties and spectacles, before becoming "England's home of mystery" in the 1870s (and responsible for inspiring Georges Méliès's interest in magic in 1884). Nevil Maskelyne and George Cooke's shows ran for thirty years, incorporating increasing amounts of film, before the hall's demolition in 1905.

ered pace when the management of the Alhambra Music Hall in Leicester Square approached Paul, proposing a fortnight's run as part of the theater's variety bill—clearly designed to compete with the Lumière Cinématographe, which had opened on 9 March at the Alhambra's Leicester Square rival, the Empire, presented by the multitalented entertainer, "Professor" Felicien Trewey. The Alhambra engagement, for which Paul's projector was renamed the "Animatograph," began on 25 March and would famously run far beyond its initial booking.[57] Paul was now committed to presenting a ten-minute program six nights per week, for the substantial daily fee of £11, or £66 per week. Still apparently believing that interest in moving pictures might prove short-lived, he made arrangements to present a similar program at "five or six other [music] halls—the Canterbury in Westminster Bridge Road, the Paragon, the West London, the Britannia at Hoxton, as well as one or two others."[58] To maintain this hectic schedule, Paul hired a brougham carriage to travel his growing circuit and began to recruit operators: "Mostly I hired lime-light men from the music halls. I paid them the hitherto unheard of salary of four pounds a week to project pictures."[59]

Projected pictures had become an established attraction in a matter of weeks, and not only in London. The Cinématographe was attracting crowds in Paris, while in New York, Edison had bowed to the popularity of projection over the Kinetoscope and acquired an existing projector, Thomas Armatt's Phantascope, renaming it in the process. The launch of "Edison's Vitascope" took place on 23 April at Koster and Bials' music hall on Broadway, with a program of Kinetoscope subjects supplemented by the established Acres-Paul success *Rough Sea at Dover*, which attracted enthusiastic praise, with no reference to its source.

We have an account of the rival operations in Leicester Square at this time by another interested party. Charles Webster had arrived in London on 30 April, representing Edison's licensees, Raff and Gammon, who were hoping to launch the Vitascope in this film-mad city. Webster went first to the Empire to assess the competition:

> I must say I was surprised to see such good results. . . . They have been at the Empire for two months and exhibit afternoons on a 50% basis, and evenings they get £10. Mr C[inquevalli, Edison's agent already in London] tells me it runs close to $600.00 a week and it makes the hit of the show. In operation it is noiseless, being operated by hand power, and takes but about 2 sq ft. The wait between pictures is between 15 and 20 seconds. They use the transparent screen and are located about 20 ft from it, being on the stage). The picture is about ½ as bright as ours, and the picture a trifle smaller.[60]

He went on to describe the program in detail, observing that the films "are of a local nature and all true to life . . . and cost a mere nothing compared to ours." He also noted:

> There are two or more other machines in London, making three in all, but the one at the Empire is the best of all. . . . PS—Have just found out that the other machines are sold for $200.00 and that quite a number have been sold to France.[61]

The "other machines" were clearly Paul's, at the Alhambra and Olympia. Perhaps the rest of Paul's circuit had not yet started, or Webster had failed to inquire more widely. But his account points to several important issues regarding Paul's equipment.

16. Exterior of the Alhambra in Leicester Square, with large signs advertising Paul's Animatograph (1896).

The first is perhaps the question of quality: how good was Paul's original projector? According to Wood:

> [Paul] confessed to me that he saw the Lumière pictures at the first opportunity and considered them better than his own because they were steadier and brighter.[62]

Other reports from this period confirm the general verdict that the Cinématographe produced better results than the early Theatrograph.[63] But Paul was committed to improving his device and began to modify his demonstration model almost immediately. His first projector, patented on 21 February, used a seven-pointed star-wheel for its intermittent mechanism. Paul explained the reason for this when writing to his fellow pioneer Messter, who had asked about details of early mechanisms:

> The reason I first used a seven pointed intermittent motion was this: the Kinetoscope had a fourteen picture sprocket, which had evidently too much inertia for an intermittent action, so I made the sprocket half that diameter, of aluminium and as light as possible.[64]

Paul's second patent was dated 2 March, sixteen days before that for Jules Carpentier's five-pointed star wheel and before a further patent by Victor Continsouza that used the four-pointed "Maltese Cross."[65] Most later projectors by Paul and others would use three- or four-pointed crosses, "with tangential action replacing the primitive form."[66] As for noise, Paul affirmed that Messter was "correct in supposing that the mechanism was noisy," but explained that in theaters the projectors were behind the screen, "usually in a compartment at the extreme rear of the stage," so was "not very objectionable."[67]

Webster's information on price has a much wider significance than the figure quoted. Paul was in fact the *only* manufacturer in Europe selling workable projectors on the open market in March–April 1896 — with the inevitable result that would-be "early adopters" turned away by the Lumières, who were only willing to licence territorial franchise-holders at this stage, found their way to Paul. One of the first of these was the American magician Carl Hertz, who left an entertaining account of his struggle to buy one of Paul's few projectors for his world tour in March 1896:

Paul agreed to sell me a machine for £50, but said that he could not deliver it for two or three months. I told him that I was leaving for South Africa on the following Saturday — it was then Tuesday — and that I would like to take the machine with me. He said that he had only two machines, and that these were on the stage at the Alhambra, where he was fulfilling a six months' engagement at £100 a week. I offered him £80 but he would not listen to me, and I went away much disappointed.

The next night I called to see him again, took him out to my club to supper, and did all I could to induce him to sell me one of his machines. But it was no use; he would not do so. However, on the Friday night . . . I determined to make a last attempt, and accordingly took him out to supper again and offered him £100 for one of his machines. He repeated, however, that he could not risk parting with one; he must have a machine in reserve in case of accidents.

"Well," said I, "you had better take me over to the Alhambra and explain to me the workings of the machine and all about it, so that I shall understand it when one is sent out to me."

So we went back to the Alhambra, where he took me onto the stage and showed me the whole working of the machine. . . . We were there for over an hour, during which I kept pressing him. . . . Finally, I said: "Look here! I am going to take one of the machines with me now."

With that I took out £100 in notes, put them in his hand, got a screwdriver, and almost before he knew it, I had one of the machines unscrewed

17. Carl Hertz was the stage name of Louis Morgenstein (1859–1924), a successful international magician who took Paul's projector on his round-the-world tour of 1896–1898, giving debut film shows in many cities throughout South Africa, India, Australia, New Zealand, and subsequently Singapore and Burma.

from the floor of the stage and onto a four-wheeler. The next day I sailed for South Africa on the Norman, with the first cinematograph which had ever left England.[68]

If this story is substantially true—and Hertz certainly did leave with a projector and some films (of which more in chapter 6)—then his first meeting with Paul would have been on the opening day of the Alhambra engagement, Tuesday, 24 March. This fits with Will Day's account of how the unplanned sale to Hertz led to an improvement in the projector's design. According to Day, Paul "deeply regretted" allowing himself to be "cajoled" by Hertz

as it necessitated him working without any rest from the Saturday to the Monday evening to produce another machine. [However] this proved to be a blessing in disguise [since it] caused Mr Paul to introduce an improvement to the Animatograph, by the introduction of a third sprocket, running continuously, to feed the film from the top spool to the intermittent sprocket.[69]

Paul's own account of the projector's evolution stated that "my first sale of a projector was in March 1896 to [the magician] David Devant; and it resulted from a descriptive article in the *English Mechanic* of March 6th 1896."[70] However, he also noted that "two only of this type" were made, one having been the original demonstration machine in February. So the model Hertz bought was likely one of the two dozen "suited for short films only" that Paul produced before switching to a "top feed" model, which he then "made in batches of 24 or 36 at a time." Day implies that he is quoting from Paul's sales ledger when he lists Esme Collings of Brighton as having bought a No. 2 model on 17 March. The other purchasers he lists are A. D. Thomas (4 April), Mr. Howard ("who first worked this apparatus on the Moss and Thornton circuit," also on 4 April), and Mr. Méliès of Paris, "who is recorded as having purchased 6 Theatrograph projectors on 4th April." Méliès then proceeded to "encase one of these instruments in a box and [convert] it into a motion picture camera."[71] Another early cross-channel customer noted by Day was Charles Pathé, the future tycoon of early cinema, and already one of Paul's Kinetoscope customers.[72]

Although anecdotal and documentary evidence cannot be fully reconciled, Paul's willingness to sell Theatrographs (often known as Animatographs, after the name adopted for the Alhambra) meant that demand far outstripped what he could produce, even though he could virtually name his own price. As we shall see, it was significant that at least three of his early customers were magicians; quick to see the potential of this new machine, they would have an important influence on the direction that early moving-picture entertainment took.

Claims and counterclaims about priority flew within the photographic press throughout the first half of 1896, as the commercial appeal of moving pictures became clear. To Acres's letter of 7 March, claiming Paul knew nothing about the principles involved in creating a moving-picture camera, Paul replied ten days later that it was he who had supplied Acres with a complete "kinetograph," "in designing the mechanism of which [Acres] took no part."[73] Henry Short was invoked by both in the argument, and after Acres wrote that it was William Friese-Greene who "holds what is really the master patent,"[74] Friese-Greene himself joined in to recall his patents of 1890 and 1893 "for producing so-called animated photographs upon a screen,"[75] now adding a new claim—which would result in his

18. A Paul Theatrograph acquired by Georges Méliès early in 1896, now in the collection of the Cinémathèque Française.

continuing to be venerated as England's "inventor of kinematography" until the late twentieth century.[76]

According to Day—by no means always reliable, but one of the first to take an historical interest in early film equipment—it was the production of the No. 2 Theatrograph that led to Paul constructing a new camera sometime in March 1896. With this he would enter film production on his own account for the first time, to meet the demand among his growing number of Theatrograph customers. And immediately after the bustling London scenes that were his first subjects came two "rough sea" films, taken at Ramsgate, as if to match or trump Acres's pioneering *Rough Sea at Dover.*

CHAPTER 3

"Adding Interest to Wonder"

THE FIRST YEAR IN FILM

By April 1896, Paul faced a unique range of opportunities and challenges. His two London exhibition engagements were delivering substantial weekly receipts from Olympia and the Alhambra.[1] This had been achieved without producing any new films since the agreement with Acres had ended a year earlier. Having created his own projector, Paul was one of the very few suppliers selling moving-picture equipment in Europe, and he was barely able to meet the high level of demand.

We must wonder how Paul coped with this dramatic increase and change in direction for his instrument-making business. There are stories of customers camped out on the stairs of his Hatton Garden workshop, and the cab was on duty nightly to carry him from show to show—but surely he also had to take on new staff and new facilities? We have no information about numbers of employees, other than the projectionists he had trained, but know that at some point he rented additional premises in Leather Lane, which became in effect his "Animatographe department."[2] Yet even as he sought to cope with, and capitalize on, the surge of demand, Paul faced great uncertainty. Like Edison and the Lumières, he had no way of knowing how long excitement over projected pictures would last—or how best to prolong it, except by throwing himself wholeheartedly into the new business. The fifteen months between May 1896 and August 1897 would see him produce, tour, and innovate on an extraordinary scale, more intensively than any of the other pioneers.[3] And amid this frantic activity, he would experiment with new kinds of subject, while bringing the experience of moving pictures to a vast range of audiences, throughout Britain and around the world.

Producing Pictures

Like his original manufacture of moving-picture equipment, Paul's start in production came in response to a proposal, which he quickly adopted. The Theatrograph had been launched without any new films, and he had had no camera since his split with Acres. But modifying his first projector had led to the creation of a new camera, and by April, with his audiences and outlets in London multiplying fast, there was an urgent need for new subjects. Apparently it was Alfred Moul, the manager of the Alhambra music hall, who "wisely foresaw the need for adding interest to wonder and staged on the [Alhambra] roof a comic scene called 'The Soldier's Courtship,' the 80 ft film of which caused great merriment."[4] Moul's choice of subject was no doubt prompted by his theatrical experience. John Poole's farce *A Soldier's Courtship* had been a mainstay of English popular theater since the 1820s. The "Adelphi screamers," as they were known because of the London theater that commissioned them, provided a solid vehicle for well-known comic actors to improvise and deliver the lively physical action that made them a vital part of the playbill.[5] In another account of how Paul came to make the film, he recalled Moul suggesting "that I should make a short comedy in order to put a few laughs into the program of scenic and interest films I was showing."[6]

The cast of three principals—soldier, fiancée, and intruder—was all drawn from the company currently appearing in the ballet *Bluebeard* at the Alhambra, which also supplied the minimal setting.[7] Fred Storey, who played the amorous soldier, was already an established stage star, having made his name in Augustus Harris's Drury Lane pantomimes, and was now a leading dancer in the Alhambra's popular ballets. The object of his affections was played by Julie Seale, also a dancer at the Alhambra. But most intriguing is the lesser-known dancer who played the busybody that intrudes on the courting couple. Ellen Daws, or "Dawn," as she was known on stage, would become Robert Paul's wife in August of the following year (about which more later).

According to Leslie Wood, Moul directed the actors on the roof of the Alhambra, while Paul operated the camera.[8] The roof of a city-center theater would also serve as a stage in New York, where Frank Vincent and William Paley filmed a Passion play at the end of the following year, initially passing it off to showmen and their innocent audiences as the celebrated version staged in Oberammergau, Bavaria.[9] The advantages of such locations were numerous. Since theaters were then among the highest buildings in the city, unwanted urban background was eliminated. Rooftops ensured maximum available daylight without intruding shad-

19. Proposed by the Alhambra's manager, Alfred Moul, *The Soldier's Courtship* (April 1896) starred two of the theater's leading dancers, Fred Storey and Julie Seale, with Paul's future wife, Ellen Daws, in a "character" role interrupting their courtship.

ows, which suited the first cameras and their comparatively "slow" film stock.[10] And being on top of a theater afforded easy access to its store of props and costumes. In *The Soldier's Courtship*, the park bench is schematic, but Storey and Seale are smartly costumed, the latter with a stylish feather, and Daws is apparently padded out to make her a "lady of mature years," as noted in the *Era* review. But since this almost legendary film has been found and restored, it is the enthusiastic yet graceful performances by three skilled dancers that impress.[11] Little wonder it proved popular — and was remade as *Tommy Atkins in the Park* two years later in the rural setting of Muswell Hill, where Paul was building a new studio.

While *The Soldier's Courtship* was indeed Britain's first staged or "story" film made for screening, the Paul-Acres partnership had produced *Arrest of a Pickpocket* in 1895, for showing on the Kinetoscope, and Paul would produce another acted film in mid-1896, *Arrest of a Bookmaker*, in which the police apparently struggle to hold a street bookmaker.[12]

Why was it that both of these pioneers — and in particular Acres, who was strident in asserting that the new medium should not be used for trivial purposes — chose to portray such violence? No other pioneer filmmakers, in America or France, would venture to show scenes of robbery, fighting, and arrest until around 1900, after which, of course, they became commonplace in cinema.[13] Yet in Paul's production of late 1897, we find not only the knockabout struggles of *The Miller and the Sweep* and *The Rival Bill-Stickers*, but a series of films based on fights: *A Lively*

Dispute, Theft (in which two tramps steal a goose and are pursued by a farmer's wife), *Quarrelsome Neighbours, Robbery,* and *Jealousy.* The only one of these extant, *Robbery,* is also known as "a wayfarer compelled to disrobe partially," which describes exactly what happens as an aggressive figure with a revolver confronts a bowler-hatted man. *Jealousy* apparently showed "Dramatic scene in gardens — jealous husband shot."[14]

The prevalence of violent scenes may help to explain why the first historian of British cinema, Rachael Low, compared "Paul's taste" unfavorably to that of "the other chief directors of the period . . . [who] do not seem to have made so much game out of marital strife, drunkenness, absence, physical cruelty and other themes recurrent in Paul's work."[15] Contrary to Low's surprisingly prim judgment, the social historian Paul Langford observed that the English had a long-standing reputation, especially among their continental neighbors, for aggression and readiness to resort to physical violence.[16] Yet, he noted:

> Even the English footpad, it was said, was rarely the murderous thug that he would have been on the Continent. Defoe described an "English way of Robbing generously, as they call it, without Murthering or Wounding."[17]

Paul's earliest fictional films seem to confirm this, but so too do those of his contemporary James Williamson (*Stop Thief,* 1901) and his onetime employee Frank Mottershaw, with his much-imitated *Daring Daylight Burglary* (1903).[18]

It would appear that Paul was keen to assert his independence after the failed partnership, with films that eclipsed Acres's earlier successes, *Rough Sea at Dover* and the climax of the 1895 *Derby.* And Paul's 1896 *Derby* film would unexpectedly lead to the royal patronage that played a valued part in publicizing early film. On 27 June 1896 Acres had filmed a visit by the Prince and Princess of Wales to the Cardiff Exhibition, which he wanted permission to exhibit publicly.[19] According to an official account, "Before giving his permission, The Prince of Wales asked Acres to bring the film to Marlborough House for inspection."[20] That screening took place on 21 July 1896, with Acres assisted by the future producer Cecil Hepworth. There had been press reports of Acres having made a hole in the exhibition wall to gain a better view of the visitors — allegedly with permission, although not from the royal party — and in this (now lost) film, the Prince of Wales was seen scratching his head. Despite this "indiscretion," the royal couple were apparently happy to invite Acres to show the film, along with some twenty other subjects, in a marquee at Marlborough House, before forty specially invited guests.

20. Paul with his camera, posed for an article celebrating his Derby film in *Strand* magazine, August 1896.

The future Edward VII would see his horse winning the Derby on screen when he visited the Alhambra in June to see what had become the first major success of British "animated photography," Paul's coverage of the 1896 Derby, won by the Prince of Wales' horse Persimmon. The "Persimmon Derby" quickly became a news story in its own right, with articles in a variety of newspapers devoted to its making and triumphant screening. Paul seems to have realized at this very early date the potential for publicity in filming a topical subject and showing the result almost immediately. He hired a wagonette and set up his camera beside the course at Epsom in Surrey, where the Derby, one of England's premier sports attractions, had been staged since 1780. At the track, Paul later recalled, he was harassed by a gypsy bookmaker, who thought he was trying to steal a lucrative site.[21] Having lashed his cart to the fence in defiance of the gypsy and filmed the end of the race, followed by crowds swarming onto the course, Paul returned quickly to London and developed and printed the 40 feet of exposed film, which he showed triumphantly at both the Alhambra and the Canterbury on the following night.

The most vivid account of the film's reception at the Alhambra appeared in a sporting paper:

> For a few seconds we are all expectation, now the crowd is yelling "Persimmon! Persimmon" that we do not hear, but we see the finish of the race, the favourite and the winner nearly locked together, though the lead of the Prince of Wales' horse is clear. Earwig [third?] seemed a long way behind and the rest nowhere. The camera cannot lie, whatever the case may have been with George Washington. The audience went into ecstasies, and clapped its hands, some cheering the while so fervently as to draw [drown?] the strains of the orchestra playing "God Bless the Prince of Wales."[22]

According to both *The Era* and Paul's illustrated interview in the *Strand*, the film attracted such enthusiasm that it had to be reshown three times, very much in the style of the encores performed by other music-hall acts:

> An enormous audience at the Alhambra Theatre witnessed the Prince's Derby all to themselves amidst wild enthusiasm, which all but drowned the strains of "God Bless the Prince of Wales," as played by the splendid orchestra.[23]

Like anecdotes of early spectators cowering before the Lumières' *Arrival of a Train at La Ciotat*, although with considerably more authentication, the "Persimmon Derby" entered the mythology of early cinema in Britain.[24]

Clearly, the sense of immediacy played a part in its success: news of the race result was still fresh when the film appeared. That the race was won by the Prince of Wales' horse, rather than the favorite, meant the film benefited from widespread enthusiasm for this most "sporting" member of the royal family, an association further enhanced by the apt musical accompaniment. Mounting a live horse race on stage was not beyond the ingenuity of the late Victorian theater, as proved by the stage spectacle of *Ben-Hur* and its successors.[25] But the few flickering seconds of Paul's film made an enduring impact and created a story that helped define Paul as a particularly *English* pioneer, especially after the Prince of Wales came to the theater to relive his famous victory.[26]

Royal patronage or association was important for the reputation of

Marvellous
ANIMATED PHOTOS.
PAUL'S

The Great
ALHAMBRA SUCCESS
Original

THEATROGRAPH

COMIC,	**SCENES**
Military,	of
Naval,	London
Domestic	Life.
and	
Topical	Selections
Scenes.	from
	.50
Constant	Stirring
Change.	Subjects.

The Prince's Derby will be run Daily
(See "Strand Magazine" for August)
ss Maud's Wedding &c. &c.

frs to a pitch of excitement.	*Times.*—Rapturously received and repeated.
up and wonder.	*Lloyds.*—Nothing more realistic was ever devised.
raw a thundering demand for an encore.	*Daily Mail*—Received with yells of enthusiasm.

21. Paul was quick to capitalize on the popularity of his Derby film in print advertising.

this new entertainment, especially in Britain, and just as Acres prized his presentation "by royal request" at Marlborough House, so Paul would later recall how his early customers included representatives of the Danish and Swedish royal families. And within a year, he would play a prominent part in making Queen Victoria's Diamond Jubilee the first successful royal film event on a global scale.[27]

Modern Magic

Paul's earliest customers included a number of well-known magicians, including the leading English conjurors David Devant and Nevil Maskelyne and others from abroad, as already noted, Carl Hertz and Georges Méliès. The reason for their interest in the new apparatus may not be immediately obvious in an era when magic performance has tended to concentrate on "close-up" illusions, apparently performed without apparatus, even though television has contributed to reviving the nineteenth-

22. David Devant (1868–1941; *left*) was the first English magician to become involved with Paul, performing in a number of his early films and presenting them in the interval of the Promenade Concerts at the Queen's Hall in 1896. Nevil Maskelyne (1863–1924; *right*) had a long-running magic show at the Egyptian Hall and managed touring magicians. Also an inventor and radio pioneer, he appeared in Paul's films and was listed as a subscriber in the 1897 company flotation.

century tradition of magic as dramatic spectacle. Magic performances had become an important theatrical genre during the later nineteenth century thanks largely to magicians' long tradition of exploiting new and relatively unknown technologies to astonish audiences—a tradition that played an important part in the development of science itself, with ancient "magical" machines seen the forerunners of modern scientific instruments, and the lore of "natural magic," as expounded by David Brewster, effectively a course in basic science.[28] The great French magician Jean Eugène Robert-Houdin (1805–1871), whose theater Méliès had reopened as an act of homage—in the same spirit as Erich Weisz taking the stage name "Houdini"—used electromagnets for some of his most famous illusions and took an interest in constructing automata. With such antecedents, it was inevitable that magicians would want to exploit this latest form of illusion.

Paul's very first purchaser of a projector, David Devant, was a popular performer at the Egyptian Hall in London, who toured widely as well as performing for select society events, including royal parties. An incident recounted in his memoirs illustrates how conjuring at the turn of the twentieth century brought together new media and old concerns. Devant performed a trick with his sister in which she appeared to read cards that were in sealed envelopes. This so baffled Sir Oliver Lodge, a leading physicist who was also active as an investigator of "psychic phenomena," that he refused to believe it was only an illusion, despite assurances from both Devant and Maskelyne that "the results were obtained by trickery"—in this case, using an electric light to read the cards through the envelopes, then passing their contents to the performer by speaking tube.[29] Devant seems to have presented moving pictures regularly as part of his performances, no doubt with practiced showmanship. It would have been his prestige that led to a special presentation of Paul's films in the interval of the popular Henry Wood Promenade Concerts, then held in the Queen's Hall, Langham Place, in September 1896.[30]

Devant was quoted on the front of Paul's 1898 catalogue paying tribute to the "beauty . . . steadiness, detail and clearness" of Paul's films, which he had exhibited "continuously since March 1896."[31] Despite the claim that this was an "unsolicited testimonial," Devant had already performed some of his best-known illusions for Paul to film, as in *Devant Produces Rabbits from a Hat* and *Devant Conjuring with Eggs*. He also may have acted as an agent for Paul, probably receiving a commission on sales to fellow magicians, while Walter Booth, soon to become an important assistant to Paul, was a member of one of Devant's touring companies.

Maskelyne, with whom Devant worked closely, also featured in an 1897 Paul film, performing his plate-spinning act.[32] There must have been a continuing business relationship in this case too, since Maskelyne is listed as one of the directors of Paul's proposed Animatographe Company in March 1897. And, like Méliès's Théâtre Robert-Houdin, Maskelyne's Egyptian Hall would gradually increase its proportion of film presentation to live performance. Maskelyne is largely remembered today for his strenuous campaign to expose the use of illusionism by such bogus spiritualists as the American Davenport brothers, who toured Britain in the 1860s and became a popular target for debunking by fellow magicians.[33] Apart from providing a "narrative' for stage magic, this may have been inspired by an ethical impulse to undeceive or to challenge the gullibility that surrounded the Victorian culture of spiritualism. Robert Browning had already attacked the bogus American medium Daniel Home in his popular dramatic poem "Mr Sludge, 'The Medium'" (1864), and Paul would later produce several films in the same vein.

Paul's comic fantasy *Upside Down; or, The Human Flies* (1899) starts with a conjuror-medium performing conventional "table turning" before he magically turns the whole room upside down in a breathtaking display of filmic legerdemain that remains impressive. This was achieved by the

23. *Upside Down* (1899), subtitled *The Human Flies,* shows a séance party who suddenly find themselves dancing on the ceiling—thanks to an ingenious splice, and a magician who has now been identified as Walter Booth, a new recruit to Paul's staff.

relatively simple means of cutting to a painted "upside down" set, against which members of the séance party happily cavort in reverse motion, as if on the ceiling. Such a polished display of conjuring skill may well have been prompted by Walter Booth joining Paul's staff at the Sydney Road studio, as he apparently plays the conjuror.[34] Much later, in *Is Spiritualism a Fraud?,* also titled *The Medium Exposed,* a fraudulent medium who has smuggled in an accomplice to act as a "spirit" is punished for his deceit by being paraded through the suburban streets tied to a cart, for all the world like a modern witch being publicly humiliated.

Although *Is Spiritualism a Fraud?* is clearly intended as a comedy, the considerable violence directed against the deceitful medium suggests a link with Maskelyne's denunciations of fake spiritualism. In both cases, showing how the spiritualists produced their effects provides a dramatic pretext for illusionism—a demonstration that is also a performance, as if magic requires some justification to rescue it from triviality, or from something more sinister. Maskelyne and Devant's book *Our Magic* proclaims the "lofty principles" that should guide the magical profession, which they insist must never become "a branch of 'show business.'"[35] For Maskelyne and Devant, the idea of the "trick" is "utterly abhorrent." They prefer "the word 'experiment,'" and claim that "modern magic . . . deals exclusively with the creation of mental impressions . . . of supernatural agency at work."[36] In pondering how a serious young engineer like Paul could have survived, and thrived, in the world of music halls and side-shows, it is easy to imagine that Maskelyne's defense of the "high art" of magic, and use of quasi-scientific terminology, might have appealed to him. And if, as Maskelyne claimed, "the average magician of today has been educated at a public school and is socially qualified to rank with members of any other profession," might not his new colleague, the animatographer, aspire to the same status?[37]

Around Britain

By the early summer of 1896, Paul was running as many as six regular film shows around London. But there was clearly demand, and poten-tial profit, in the provinces as well. Most of Britain's provincial cities had at least one music hall, the majority belonging to one of the big circuits, such as Moss Empire Theatres. Forerunners of the cinema circuits that would cover the country after the First World War, they were the first venues where moving pictures appeared as a regular attraction. Films were, in effect, another act in the program, and so were presented by showmen, who toured like any other performer on the variety circuit, or by organized networks of operators, as Paul himself had been in Lon-don. Two of Paul's earliest customers were prominent showmen, A. D. Thomas, popularly known as "Edison Thomas," and a Mr. Howard, both of whom bought Theatrographs and a stock of films on 4 April 1896.[38] The outlay required was considerable: Paul charged £75 for the projector and £3 for each 40-foot film, of which at least ten would be needed for a fifteen-minute show.[39]

Discovering exactly where Paul's projectors and films were shown is complicated by the vagueness of both showmen and newspaper re-viewers in describing what was being used. Neither the Lumière Ciné-matographe nor Edison's Vitascope was on open sale at this time, but the term "cinematograph" had caught on as a general term for moving-picture apparatus, and Edison's name was already a popular brand for all things marvelous. So, for instance, the presentation at the New Empire Palace of Varieties in Leicester from 22 June 1896 was billed as "Edison's Kinematographe," but almost certainly owed nothing to either Edison or the Lumières and may well have been supplied by Paul.[40] Other show-men would use their own names for the apparatus, such as Fred Har-vard, with his "Cenematoscope," who gave the first moving-picture show in Leicester at the Tivoli Theatre of Varieties on 15 June. Although the local reviewer reported this as a "cinematographe" show, the subjects he described — "various styles of dancing" and "a representation of the well-known boxing kangaroo" — correspond to subjects on offer from Paul.[41]

Another guide to how moving pictures were reaching far beyond London is the appearance of films taken in provincial cities — often the result of an operator being there to present a show. The Lumière operators were the most assiduous "image hunters" in this vein, with half a dozen of them contributing to the growing Lumière catalogue of "exotic scenes."[42] More prosaically, an 1898 Paul catalogue includes three Leicester scenes, under the heading "Foreign and Local Films," which were no doubt taken during one of the early Animatograph presentations in that city.[43] "Special arrangements have been made with Mr R. W. Paul to introduce the animatograph," announced the Palace Theatre's advertisement for its pantomime in December 1896, with animated photographs replacing the "usual transformation scene" in *The Forty Thieves*.[44] By the following December, the Animatograph was described as "no novelty in Leicester," but its appearance in the Royal Opera House's pantomime, *Robinson*, was welcomed, thanks to the "constant supply of up-to-date pictures and local scenes."[45]

Paul himself was active as a traveling exhibitor and cameraman as early as July 1896, and his first recorded location was significant. Brighton has had a long history as a seaside and entertainment resort, close enough to London for a day-trip, or even for commuting, once the railway arrived. Resorts, associated with pleasure-seeking and, like the pilgrimage and spa destinations that preceded them, with novelty, provided receptive audiences for new forms of entertainment. Brighton would soon have its own group of innovative local filmmakers — one of whom, Esme Collings, had been an early Paul customer — but on 6 July 1896 Paul launched an exclusive film presentation that seems to have been modeled on the Olympia show. Billed as "R W Paul's Celebrated Animatographe," its venue was the Victoria Hall (previously known as the Pandora Gallery), and screenings ran from 11:30 a.m. to 10:30 p.m.[46] The basic admission price was the same as at Olympia, 6 pence, but the offer of reserved seating at 1 shilling (12 pence) suggests that Paul or his local operator was aware of the possibility of a more refined audience in a resort long favored by the wealthy, and even royalty, as well as simple London pleasure-seekers.

There is no way of knowing how long Paul spent in Brighton, although the show was clearly branded with his name, but he found time to make a film based on typical holiday activities, which entered the growing catalogue of new subjects as *On Brighton Beach*.[47] The formula would be much used during the early months of production: a typical scene, in this case a small boat landing on the beach, is dramatized by adding some staged comic business — here, several of the passengers falling into the sea. It is likely that such films were first shown near where they were made, providing a sense of immediacy for the local audience. Some would justify wider distribution, such as *Brighton Beach* and several films from a group shot at another popular seaside resort, Douglas on the Isle of Man, but the majority seem to have rapidly been classified in Paul's catalogue as second-class subjects, "not recommended for general exhibition."[48]

The arrival of Paul's films in another provincial town, Cheltenham, was documented by Paul himself, in the published version of his 1936 memoir, and again in the first volume of Rachael Low's *History of the British Film* (1948).[49] The first appearance was a series of screenings held as fundraisers for the Cheltenham Cricket Club, between Tuesday and Saturday, 1–5 December 1896, featuring many of the films then being shown at the Alhambra. The shows were promoted with advance articles in local papers and enthusiastically reviewed:

> The chief attraction of the performances were [*sic*] the Animated Photographs which excited the surprise of the audience and gave them the pleasure which a novelty affords.

The review described *Sea Cave in Galicia* — "The sea tumbles and foams and as the angry waves sweep in, the rocky promontories break them into sheets of feathery spray" — and noted "the mirth-provoking pantomime of *The Soldier's Courtship*."[50]

Musical accompaniment was provided by the town band and special shows were put on for pupils of Cheltenham's two prestigious schools, the Ladies' and Boys' Colleges, with the famous Lady Principal of the former, Miss Beale, "publicly express[ing] her thanks to the organizers of an entertainment which had afforded so much amusement and delight." Later in the same month, a large and "all new" selection of Paul's films returned to Cheltenham as part of an entertainment presented by "Lieut Walter Cole and his Merry Folk," which also involved ventriloquism and automata.

While shows using Paul equipment and films took place in all the main cities and began to reach smaller towns during the summer and autumn of 1896, another form of circulation was also under way, based on ancient tradition.[51] Just as holiday resorts evolved from pilgrimage destinations, so fairs had their origins in the medieval world, usually linked to the religious calendar. During the nineteenth century they had also become a focus for displaying new technology. Applications of steam power and electricity were established features by the later part of the century, and fair showmen were well used to seizing on new inventions, such as X-rays and moving pictures, and finding spectacular and entertaining ways to market them to their customers. The Phonograph and the Kinetoscope were quickly adopted by novelty-seeking fair operators, as noted in a *Times* report on one of London's favorite resorts in 1895:

> Yesterday was a delightful day on Hampstead Heath, which was visited by about 100,000 holiday-makers.... That part of the heath by the railway station ... had, as usual on these occasions, all the characteristics of an old English fair brought up to date. In the Vale of Health ... a steam round-about and swingboats did a good trade, while at various places Edison's phonograph and kinetoscope were well-patronized.[52]

Many fair operators had started doing business with Paul as a Kineto-scope supplier in 1895, and would have been predisposed to continue the connection when he introduced the Theatrograph in 1896. From May of that year, Paul began advertising his films and projectors in *The Era*, a long-established theatrical periodical that also served the wider enter-tainment business; and after the *Showman* was launched in 1900, he would advertise his new "song films" as likely fairground novelties.[53]

What was emerging out of the earlier touring practices of music halls and fairs was a new breed of film exhibitor who would contract suitable venues for regular shows, making these prototype cinemas. One of the most dynamic of these was Sidney Carter, whose New Century Pictures, based at St. George's Hall in Bradford, would eventually become an im-portant film exhibition circuit in northern England.[54] A rare surviving letter in response to Carter's inquiry about a colored print provides evi-dence of how much showmen were willing to invest to enhance their pre-sentations. Paul had shown a hand-colored film as early as April 1896, when the *Evening Standard* reported "an Eastern dance, depicted on the screen in all the gorgeous colors the scene warranted."[55] That 40-foot film—which may have been Edison's belly dancer Fatima, since there is no such subject in Paul's own list—had apparently taken Mr. Doubell, responsible for the Regent Street Polytechnic's famous hand-painted lantern slides, a month to execute. In 1902 Paul's manager, Jack Smith, quoted Carter a price of 5½d per foot and a further 30 shillings per 100 feet to color *The Prodigal Son*; the total cost of the film noted on the let-ter is £15.3s.4d—the equivalent of over £1,000 today.[56] The equally rare poster for a screening at the New Drill Hall in Durham later that year shows how this translated into a headline attraction: "A splendid presen-

24. *Blackfriars Bridge*. Paul and Acres had failed to film this newest of Thames bridges (1869) in an early test of their first camera. But in April 1896, Paul made a successful film, looking north toward his former school, the City of London, and capturing a vivid cross-section of Londoners crossing.

tation of *The Prodigal Son* in Colour. The Latest Invention in Animated Photography."[57]

The Repertoire in August 1896

The summer of 1896 saw Paul experimenting with many kinds of production. These fell into three main genres: naturalistic scenes, soon known as "actualities"; arranged incidents; and filmed variety performances. Even in such short and seemingly simple scenes, there is considerable evidence of craft and considered composition, and often interesting contrasts with other pioneers' films in similar genres. What we do not know is how many of these Paul may have filmed himself.

He was of course a Londoner, living at this time in High Holborn and working in Hatton Garden, so the busy streets of the great metropolis would have been obvious subjects. Films taken around Westminster and on Blackfriars Bridge proved especially popular. The latter, in particular, revels in the variety of late Victorian street life, but also tacitly pays tribute to Paul's alma mater. As already noted, Paul no doubt chose to place the camera at the south end of the bridge, looking north toward the Victoria

25. *Hyde Park Bicycling Scene.* Cycling was becoming increasingly popular in the 1890s, especially among women, and this view of Hyde Park includes a notable proportion of unaccompanied women cyclists.

Embankment, such that the spires and steeply pitched roof of the City of London School can be dimly seen as background to the passing traffic and pedestrians. Both bridge and building were relatively new when the film was made. The rapid growth of railway traffic had made a new bridge necessary by midcentury, and Thomas Cubitt's originally matched rail and road bridges, with their distinctive Venetian Gothic–style ironwork, were opened by Queen Victoria in 1869. The City of London School had been based in Milk Street in the City since 1837, before obtaining this high-profile site on the new embankment; the French chateau–style building opened in 1883, with Paul in its first cohort of pupils.[58]

Also popular were scenes of relaxation and entertainment, including *Hampstead Heath on Bank Holiday, Hyde Park Bicycling Scene,* showing the new craze for cycling among women, and *A Comic Costume Race,* taken at the Music Hall Sports, an annual charity event held at Herne Hill in South London in July. This last may have followed from Paul's growing involvement in the music-hall world, which he toured nightly by cab, supervising film shows. Like Edison, Paul saw music-hall acts as a natural subject, and among the surviving films is one of George Chirgwin (1854–1922), nicknamed "the White-Eyed Kaffir," who sang and clowned as a very unusual blackface "minstrel." With a white diamond over one eye and an immensely tall hat, or "quartered" in skin-tight black and white like a heraldic figure, Chirgwin cut a unique figure, and a contemporary review of one of Paul's two films of him spoke of his "familiar features" being "hailed with delight by gallery and pit" at the Paragon Theatre of Varieties in Mile End Road, an early building that helped launch the career of the great theater architect Frank Matcham.[59] Ten years after

Paul's Chirgwin film played at the Paragon, the young Charles Chaplin would appear there as part of Fred Karno's company.

Children also appear frequently as subjects, with two informal films titled *Children at Play*, and the staged *Twins' Tea Party*, in which two little girls engage in escalating warfare—very different from the well-behaved infant of the Lumières' *Baby's Breakfast*. Not content with recapitulating Acres's 1895 film of the Henley Regatta, an annual rowing and boating event on the River Thames, Paul also made a dramatized Thames subject, *Up the River*, in which a child falling off a cruise boat is rescued by the prompt action of a passerby on the bank—a role Irene Codd maintained was played by Paul himself.

Bringing the World to Britain

At the end of August 1896 Paul was showing at the Alhambra a varied program of over twenty "animated photographs," which included subjects filmed in Glasgow and Rothesay, Scotland, and Calais, France. But already he had much greater ambitions, clearly realizing that some kind of organizing structure would improve the prevailing pattern of diverse scenes—and no doubt aware of the international scope of the Lumière programs showing next door in Leicester Square at the Empire. In September, he dispatched Henry Short to the Iberian peninsula to film a series of typical views, which would be shown two months later as *A Tour in Spain and Portugal*. In effect, Paul was following the practice initiated by the Lumières—sending operators to capture what seemed most exotic or picturesque as they traveled through unfamiliar territory—a practice already well established among lantern slide and stereograph suppliers. Since road travel within Spain and Portugal was still difficult, much of the tour relied on shipping routes, taking Short from Southampton to Gibraltar, then to Cadiz, to reach Seville and Madrid, and finally to Lisbon and its nearby coast.[60] Short followed established tourist routes, filming mostly typical scenes of the kind that would have attracted a lantern lecturer gathering material for a travelogue.[61]

Some of the fourteen films eventually selected for the series[62] were shown by Short at the Coliseu Real in Lisbon, in advance of their English premiere at the Alhambra on 22 October. There they were "received night after night with much favour"—although, *The Era* noted, "their popularity is not greater than some of the old favourites," such as *The 1896 Derby* and *The Gordon Highlanders*, shown leaving Maryhill Barracks, Glasgow.

Only two of these Iberian films are known to exist today: the lively

THE ANIMATOGRAPHE.

139th Production at the

ALHAMBRA

THEATRE OF VARIETIES

August 31st, 1896.

Invented and Produced by Mr. ROBT. W. PAUL, M.I.E.E.

The series of Animated Photographs presented this evening are:

1. Scene on the River Thames. showing the rescue of a child from drowning, July, 1896.
2. Street Scene in Glasgow. August, 1896.
3. Engineers and Blacksmiths at work at Nelson Dock.
4. Passengers disembarking at Rothesay Pier, N.B.
5. The Arrest of a Bookmaker.
6. Children at play in the Archbishop's field, Lambeth.
7. Westminster Bridge Approach, near Guy's Hospital.
8. Party landing from a boat on Brighton Beach.
9. Traffic on Blackfriars Bridge.
10. The arrival of the Paris Express at Calais (Maritime).
11. Comic Costume Race at the Music Hall Sports. 1896.
12. Princess Maud's Wedding: the Processions leaving Marlborough House.
13. The S. S. "Columba" leaving Rothesay Pier, August 15th, 1896.
14. The Twins' Tea Party.
15. Henley Regatta, 1896. View on the River.
16. 2 a.m., or the Husband's Return.
17. Morris Cronin, Club Manipulator.
18. The Soldier's Courtship.
19. The Prince's Derby, 1896.
20. The 92nd. (Gordon) Highlanders leaving Maryhill Barracks, Glasgow, August 14th, 1896.

OVER.

26. Program of Paul's films showing at the Alhambra Theatre, August 1896. Nonfiction scenic and topical subjects predominate.

Andalusian Dance, preserved as a Filoscope, and *Sea Cave Near Lisbon,* which was immediately recognized as something of a pictorial triumph, remaining on sale for at least seven years. Filmed at Cascais, near Estoril, the film shows great waves breaking on rocks as seen from inside a cave known as the Boca da Inferno, or Hell's Mouth. It could easily have been a successful still photograph, with the mouth of the cave making a strong foreground frame. But by combining the kinetic appeal of a rough sea with a striking composition, Short had created a worthy successor to both Acres's and Paul's own breakwater scenes. The film would be

27. *Sea Cave Near Lisbon*. Part of the series *A Tour in Spain and Portugal*, filmed for Paul by Henry Short in the summer of 1896, this view of breaking waves from inside a cave at Cascais known as the Boca do Inferno (Mouth of Hell) was much admired for its scenic qualities and became the first film that Paul copyrighted.

described in Paul's first catalogues as "An artistic view of a Sea Cave,"[63] sometimes with the evocative addition, "waves violently dashing in."[64]

No doubt encouraged by the attention paid to his Iberian series, Paul sent Short on another expedition, early in 1897, to Egypt. Britain had a strong historic interest in the region, bolstered by its gaining control of the Suez Canal in 1882 and thereafter ruling Egypt as a de facto protectorate. Egypt had also been an early tourist destination for Britain's rising middle classes, with tours to Egypt and Palestine first organized by Thomas Cook in 1869. Against this historic background, Short filmed at least ten subjects, ranging from shots of the pyramids to genre scenes taken around Cairo. During the summer, Paul himself visited Sweden, ostensibly in response to the king of Sweden and Norway requiring that "the maker was to accompany the projector and see it properly installed in the palace at Stockholm."[65] But clearly this was also an opportunity to film another group of typical scenes, extending his catalogue in the way that the Lumières had with *vues exotiques*. The Swedish films did not remain in Paul's catalogue for long, and none survive, but they doubtless formed part of the Theatrograph program that enjoyed considerable success over three summer months in Stockholm.[66] However, Paul was clearly beginning to think beyond the immediate goals of expanding his catalogue and taking exhibition to a wider range of venues.

CHAPTER 4

Time Travel

FILM, THE PAST, AND POSTERITY

By mid-1897, Paul had amply demonstrated the potential of his Animato-graph to offer audiences a novel experience of "space travel," bringing the distant close with a vividness that would soon challenge even the established repertoires of magic lantern and stereoscope images. One clue as to what he thought of this new business that was consuming so much of his time and energy must lie in a strange episode from 1895 that would become his passport to later fame, long after his pioneering film work was forgotten. In October of that year, although already busy with his Kinetoscope business, Paul registered a preliminary patent for "A Novel form of Exhibition or Entertainment, Means for Presenting the Same."[1] The novelty was to give customers the illusion that they were traveling through time—an idea frankly inspired by the sensation H. G. Wells's story *The Time Machine* had caused when it ran as a serial early in the year, before appearing as a novel in May.

What Paul proposed can most easily be understood as an anticipation of today's flight-simulation attractions, common in theme parks and high-tech museums. Spectators would take their seats on one or more semi-enclosed platforms, which would be suspended from cranks, with a rack or cam mechanism beneath. The description is suitably vague, but the intention is clear: viewers will feel a "rocking motion" as the platform travels "bodily forward," while air is blown over them "to represent the means of propulsion." All this will take place in low light or darkness. Then the motion will stop, and "a scene upon the screen will come gradually into view of the spectators, increasing in size and distinctness from a small vista, until the figures may appear lifelike." The figures, "made up characters performing on a stage," would be superimposed on a separately projected landscape and could be filmed "with or without a suitable background blending with the main landscape." Landscape effects would be a combination of photographic and painted, with "changeable coloured, darkened or perforated slides" to represent "sunlight, darkness, moonlight, rain."

Audiences would have the sensation of traversing these scenes as if in "a navigable balloon" (recalling Jules Verne's first success, with the fantastic aerial travelogue of *Five Weeks in a Balloon*). The image would change size thanks to projectors mounted "on trolleys, upon rails," which was in fact a technique used in the eighteenth-century Phantasmagoria lantern shows. To complete the illusion, after visiting "a certain number of scenes from a hypothetical future," spectators would be led from their platforms to visit "grounds or buildings arranged to represent exactly one of the epochs" on the program. They would then resume their seats for a return to the present, which Paul suggested might be dramatized by a contrived "overshoot" into an earlier or later period, "before the re-arrival [is] notified by the representation on the screen of the place at which the exhibition is held," which "can be made to increase gradually in size as if approaching the spectator."

The "novel entertainment" was never launched, for many obvious reasons, but the vision it expressed has if anything become more potent and suggestive with the passage of time. Even for its proposers, Paul and Wells, both ambitious young men at the start of their careers in 1895, it would seem remarkable as they looked back on it at the end of their lives.[2]

Paul, we must remember, was still manufacturing peep-show Kinetoscopes, although the patent application refers to "powerful lanterns throwing portions of the picture … upon a screen," and to an all-important shift from the continuous film movement of the Kinetoscope to the intermittent travel necessary for efficient projection. He was also already thinking about the economic limitations of the single-viewer Kinetoscope. A year later, he said that the success of his Kinetoscopes at the *Empire of India* exhibition had made him wonder "if their fascinating pictures could be reproduced on a screen, so that thousands might see them at one time."[3] But in another interview, just six months after the patent application, it was Wells's "weird romance 'The Time Machine' [that] had suggested an entertainment to him, of which animated photographs formed an essential part."[4]

What has been missing in most interpretations of this overdetermined moment is the recognition that Paul's engagement with Wells was not one-sided, with Paul merely seeking to capitalize on Wells's imaginative coup. The technology of moving pictures was in transition, and by the end of October 1895, Paul was working on his first projector, which the "Time Machine" patent had implied. He knew by then of the Lumière brothers' Cinématographe, featured in *Nature* in August, and corresponded with them in late November, when the *Amateur Photographer* also reported on their work.[5] What he had deduced from the India exhibition—that single-viewer machines were intrinsically less lucrative than an apparatus that could be simultaneously experienced by many—would have pointed away from the elaborate requirements of the Time Machine auditorium, toward marketable projectors. Hence, as he would later say, "the time was not ripe." But for Wells, *The Time Machine* was only the first in a series of powerful "scientific romances" that would shape the twentieth-century imagination, and these owed much to the rapid development of moving pictures, which Paul was leading in Britain.

The first historian to focus on the Paul-Wells project in detail was Terry Ramsaye, who in his ambitious "history of the motion picture through 1925," had no doubt that it anticipated the developed "photoplay" of the 1920s:

> No writer or dramatist has since made so bold a gesture with reference to the screen as resulted from this tentative joining of Paul's invention and Wells's fancy.[6]

Ramsaye was well placed to survey cinema's then short history, as editor of the main American trade journal, the *Motion Picture Herald*. But he was not content merely to record the facts, and his colorful account did much to establish what has come to be accepted as the ancestry of cinema, tracing the emergence of what he calls "the Wish of the world" from prehistoric picture-making through shadowplay to the magic lantern and the nineteenth-century developments that led to Edison's breakthrough.[7] His chapter on Paul comes between accounts of the first workable American projector and of the Lumières' "sixteen-a-second" Cinématographe, and contains one of Ramsaye's most extravagant poetic outbursts: he speaks of giving "the spectator the possession of *Was* and *To Be* along with *Is*," and suggests that the "motion picture Time-machine was artistically at one with Ouspensky's mathematically mysterious philosophy and Einstein's philosophically mysterious mathematics."

While this may seem pretentious from a modern perspective, it evokes the intellectual excitement that swirled around cinema at the time Ramsaye was writing. Albert Einstein was on his way to becoming the most famous living scientist, helped by a culture of popular interest in the new physics, which would even reach film audiences via Max and Dave Fleischer's explanatory cartoon *The Einstein Theory of Relativity* (1923). And the scientific theories of relativity seemed to many somehow connected with the mysticism of theosophy and later schools of esoteric thought, such as Gurdjieff's and Ouspensky's "fourth way" or J. W. Dunne's "experiments with time."[8] Popular literature, and increasingly cinema, borrowed from these ideas to dramatize "time travel" in the 1920s—both Cecil B. DeMille's *The Road to Yesterday* and the first screen version of Arthur Conan Doyle's *The Lost World* appeared in 1925, while Ramsaye's book was in preparation—helping to reinforce the idea of cinema itself as a form of collective time travel, routinely transporting its audiences to distant eras.

Ramsaye was certainly influenced in his view of the Time Machine project by the formidable reputation Wells had acquired by the 1920s. A pioneer of the League of Nations, launched to ensure that the Great War would truly remain the "war to end wars" (another phrase he had coined), Wells had become the confidant of political leaders throughout the world, author of the popular *Outline of History* and of countless influential stories and novels. Little wonder that the discovery of his early collaboration with Paul lent added luster to the origins of cinema. Wells claimed he had no clear recollection of his involvement until Ramsaye contacted him to ask for details, which is scarcely surprising given the turmoil in his life in the mid-1890s. Still struggling to establish himself as a writer, with a punishing output of essays and short stories, he had recently abandoned his wife to elope with a young science student. Both suffered from tuberculosis, then known as "consumption," as well as from disapproving landladies, as they struggled to survive. After an uncomfortable summer spent outside the city in Sevenoaks, Kent, they returned to lodgings in central London with Wells still "writing away for dear life." At this point one of the leading editors of the time, the poet W. E. Henley—coincidentally a future neighbor of Paul's in Muswell Hill—invited him to rework an idea that had already been published in several versions, as a serial for the new monthly journal he was launching. Henley saw that the series of unsigned articles, entitled "The Chronic Argonauts," was "so full of invention" that "it must certainly make your reputation." He offered Wells advice on expanding the time traveler's journey to cover earlier periods and urged him "to tell us about the last man & his female & the ultimate degeneracy of which they are the proof."[9]

"The Time Machine" began to appear in Henley's *New Review* early in 1895, and soon attracted attention. W. T. Stead's influential *Review of Reviews* declared in March, "H. G. Wells is a man of genius," and soon the young genius found himself hired as a book reviewer and a theater critic, with three of his own books published in this *annus mirabilis*. Paul had recalled in 1914 that his proposed apparatus was "fully developed with some assistance from Mr Wells," and Ramsaye was able to confirm that he "wrote to Wells, who went to confer with Paul at his laboratory at 44 Hatton Garden." Will Day, the pioneer historian of cinema, who certainly knew Paul, wrote of "a long and interesting conference" between the two, and of "several other talks between them."[10]

The meetings may have left little immediate impression on either of the men, born within three years of each other and busy establishing their very different careers. But the fact that Paul had read and been impressed by Wells's story underlines that both belonged to a generation shaped by the new scientific education that was transforming late Victorian Britain. Compared with the apparently rising fortunes of Paul's family, Wells had early experience of genteel poverty, as his mother struggled to keep up appearances in a failing shop in Bromley. Young "Bertie" Wells escaped from the suffocating prospect of apprenticeship in a provincial drapery shop to schoolteaching, and a scholarship took him to the new Normal School of Science in South Kensington in 1884 — like the Finsbury Technical College that Paul attended, a product of the realization that Britain was falling behind in scientific education.[11] The Normal School had been established by "Darwin's bulldog," the redoubtable T. H. Huxley, to train science teachers, and although most paid fees, Wells was one of the scholarship boys. This contact between the ambitious young provincial and the elderly Huxley, who had helped popularize Darwin's evolutionary theory but since become increasingly pessimistic about mankind's future, was to prove crucial. Although Wells would barely scrape through his later courses in physics and geology, he was inspired by Huxley's vision of mankind having the knowledge and responsibility to shape its future destiny. He gave a paper at the school's debating society in 1886 boldly entitled "The Past and Present of the Human Race," and his contributions to its journal in the following year included "A Tale of the Twentieth Century" and the first version of what would become *The Time Machine.*

It seems probable that experience of the Kinetoscope influenced Wells's evocation of time travel. Ramsaye certainly believed it had, citing the book's third chapter, in which the traveler describes the sensation of setting off in time, as "the laboratory got hazy and went dark," with his servant Mrs. Watchett seeming "to shoot across the room like a rocket." For Ramsaye, this confirmed

the motion picture influence at the bottom of Wells' concept. The operation of the Time Traveller was very like the start of the peep-show Kinetoscope, and the optical effect experienced by the fictional adventurer was identical with that experienced in viewing a speeding film.[12]

More specifically, Ramsaye believed another passage in Wells confirmed that he must have seen a Kinetoscope running backward, in what was a popular early "novelty effect":

> Mrs Watchett had walked across the room, travelling, it seems to me, like a rocket. As I returned, I passed again across the minute that she traversed the laboratory. But now every motion appeared to be the direct inverse of her previous one.

Ramsaye quotes a fragment of "an editorial from the *St Louis Post Dispatch*," which compares the effect of the Kinetoscope running backward with an earlier scientific romance, the French astronomer Camille Flammarion's *Lumen*, in which "spiritual beings" are able to travel "in the contrary direction" along a ray of light, and so "witness the events of history reversed."[13] "It now seems," the editorial continued, "that the kinetoscope is to make this wondrous vision possible to us . . . the effect is said to be almost miraculous."

For Ramsaye, as an informed yet impatient industry commentator in the mid-1920s, the Paul-Wells project marked the first time a "professional re-creator of events [had] come into contact with the motion picture," anticipating the freedom of expression that cinema was still groping toward "by tedious evolution," after decades of subservience to the "intrenched motion picture business."[14] If only, he lamented, Wells had "given attention to the motion picture and its larger opportunity over time, he might have set the screen's progress forward many a year." Yet a closer reading of Wells's stories over the following years indicates that he *had* taken on board the implications of the full array of modern media technology. Indeed, as Andrew Shail has observed, Wells had, by 1900, mentioned moving pictures "more frequently than any other turn of the century author."[15] And in two extended time-travel stories of 1898–1899, *A Story of the Days to Come* and *When the Sleeper Wakes*, his projected future finds phonographs replacing print and moving pictures providing both public and personal entertainment.

In the Regent Street of "days to come," now "a street of moving platforms and nearly eight hundred feet wide," known as Nineteenth Way, the Susannah Hat Syndicate

> projected a vast façade upon the outer way, sending out . . . an overlapping series of huge white glass screens, on which gigantic animated pictures of the faces of well-known beautiful living women wearing novelties in hats

28. *When the Sleeper Wakes*, H. G. Wells's speculative novel about a Victorian man who awakes from a coma in 2100, appeared as an illustrated serial in both London (*The Graphic*) and New York (*Harper's Weekly*), January–May 1899. Part of its remarkably comprehensive vision of the future world was a prediction of massive public screens with moving-image advertising in the London of 2100.

were thrown. A dense crowd was always collected in the stationary central way watching the kinematograph which displayed the changing fashion.[16]

This was accompanied by "a broadside of giant phonographs" that drowned all conversation with their exhortations to "buy the girl a hat." Massive audiovisual advertising displays were also featured in Wells's more substantial "tale of the twenty-second century," *When the Sleeper Wakes*, which projected its nineteenth-century protagonist Graham into a glass-roofed world fraught with class antagonism.[17] Serialized in the *Graphic* during 1898, the story was accompanied by remarkable illustrations by the French artist Henri Lanos portraying much audiovisual apparatus. Wells could conceivably have seen Paul's 1897 prospectus for the flotation of his Animatograph company, which predicted the poten-

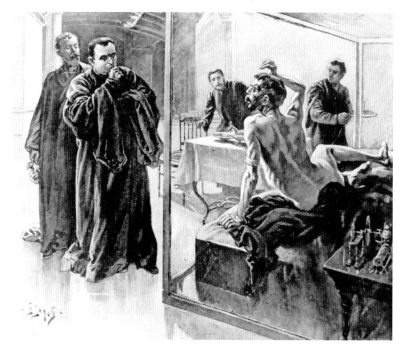

29. Significantly, Wells's awakened sleeper is shown his new costume on a small "kinetoscope-like" device, recalling Paul and Wells's discussion of a "time machine" in 1895 (*Graphic* illustration by Henri Lanos).

tial of film in advertising.[18] In any case, he foresaw the vast illuminated displays that have indeed emerged in our era of digital billboards.

He also foresaw the domestication of audiovisual media. In *When the Sleeper Wakes*, the awakened Graham is visited by a tailor who shows him a range of sartorial options by means of "a little appliance the size and appearance of a keyless watch," on which "a little figure in white appeared kinetoscope fashion on the dial, walking and turning."[19] The actual Kinetoscope was a bulky apparatus; this portable device seems closer to, if not today's computerlike watches, at least the handheld Filoscope flip-book Henry Short had patented in 1898, making use of Paul's films for its contents. While we do not know how closely Lanos was briefed, the device he pictures vaguely beside Graham's bed is described in the text as having a cylinder that implies a combination of the Phonograph cylinder with the Phenakisticope used by Muybridge and the French physiologist Etienne-Jules Marey. As with the tailor's device, Wells's confident prose conveys his conviction that such marvels will become commonplace, while the illustration can only gesture toward how they might look.

Later, during Graham's confinement by the ruling council, he discovers near his bed a tabletop apparatus with an electrical switch. When this is pressed

> he became aware of voices and music, and noticed a play of colour on the smooth front face.... On the flat surface was now a little picture, very vividly coloured, and in this picture were figures that moved. Not only did they move, but they were conversing in small, clear voices. It was exactly like reality viewed through an inverted opera glass and heard through a long tube.

Graham soon realizes he is watching a drama that refers to himself, with a joking reference to "when the Sleeper wakes." He discovers how to change the cylinders in the apparatus, and realizes he has access to a library of dramatizations. Here Wells teases his readers of 1899 by citing Rudyard Kipling's 1888 story "The Man Who Would Be King" ("he remembered reading a story of the title ... one of the best stories in the world"), followed by *Heart of Darkness* and *The Madonna of the Future*, stories by his new friends Joseph Conrad and Henry James, both of which appeared in the same year the *Sleeper* was first published. Graham also discoverers "an altered version of the story of Tannhauser," in which the hero of Wagner's opera "did not go to a Venusberg but to a Pleasure City," which we are to understand is sufficiently pornographic to embarrass this time traveler, "who forgot the part played by the model in nineteenth-century art."[20]

As Graham experiences more of this future world, he discovers that recordings have entirely replaced printed texts, paintings, and performances. And, remarkably, Wells looks forward to the film and broadcasting studios of the near future:

> factories where feverishly competitive authors devised their phonograph discourses and advertisements and arranged the groupings and developments for their perpetually startling and novel kinematographic dramatic works.[21]

The extraordinary prescience of Wells's technological and social predictions was of course a major reason for his reputation as a seer, just as it would ultimately limit appreciation of his literary achievements. But what is of narrower interest here is how quickly he incorporated not only projected pictures, then known by a wide variety of names, but also their immediate forerunner, the Kinetoscope, into his imagined world of the future. This tends to confirm that he witnessed the Kinetoscope in operation during its short-lived heyday in 1895, as Ramsaye surmised from *The Time Machine*, and possibly through his contacts with Paul. It also strongly suggests that he had seen enough projected pictures by late 1897, when he was writing *The Sleeper Wakes*, to inspire this work's bold predictions.

Such rapid assimilation and extrapolation was not altogether unusual among inventors and writers in this period. At a time when the Kinetoscope was still a peep show and the Phonograph could barely fill a medium-size room with sound, Edison confidently predicted that the combined Phonograph and Kinetoscope would "enable grand opera to be given at the Metropolitan Opera House in New York with artists and musicians long since dead."[22] Here he was following a vein of fantasy that his own Phonograph had inspired. In France, the pessimistic aesthete Villiers de l'Isle Adam and the popular novelist Jules Verne both made use of Edison's invention, the former in his darkly satirical novel *The Future Eve* (1886), and the latter in a techno-Gothic tale, *The Carpathian Castle* (1892).[23] Villiers imagined an artificial "android," whose voice would be supplied by concealed phonographs, convincing enough to attract the love of an obsessive English aristocrat, Lord Ewald. Meanwhile, Verne has another obsessive opera lover retreat to a remote castle, where he creates a shrine to a dead diva by means of recorded sound and her image captured in spectral form.

Although Wells hated being compared with Verne, it was the French writer who had first introduced modern technology into popular fiction and created such icons of modern menace as Captain's Nemo's *Nautilus* and Robur's Airship Destroyer.[24] Dramatizing Verne's most famous spatiotemporal adventure, *Around the World in Eighty Days*, had also been the foundation of Imre Kiralfy's fortunes. World's fair expositions, in London, Paris, and Chicago, were increasingly elaborate environments, in which visitors could imagine themselves traveling in space and time as they toured the exhibits.[25]

As a new century approached, travel in time as well as space was definitely on the agenda. Paul would soon find himself laying the foundations for a cinema that largely realized this ambition, as Ramsaye proclaimed in 1926 — albeit without commenting on Paul's own career in early cinema, which he may not have known. If we wonder how an apparently dedicated electrical engineer could so readily enter the world of the music halls, and eventually the early film industry, part of the answer must lie in the extraordinary fictive ambition expressed in patent application no. 19984, dated 24 October 1895.[26] With its combination of precise mechanical proposals — "shafts below the platforms, provided with cranks, or cams or worms, keyed eccentrically on the shaft" to impart rocking movement —

and "hypothetical landscapes" representing the future and the past, Paul's proposal amounted to a modernized synthesis of the Phantasmagoria and later immersive spectacles, such as the Panorama and the Diorama.[27] As we have noted, the Phantasmagoria showmen used lanterns mounted on rails to produce dramatic changes of scale in the projected image. Paul's proposal also borrowed from Dioramas, which featured large translucent paintings that changed subtly before their seated spectators, with groups of seats moved to reveal new vistas.[28] And in suggesting that his spectators "may be allowed to step from the platforms, and be conducted through grounds or buildings arranged to represent exactly one of the epochs through which [they are] supposed to be travelling," he may also have been inspired by some of the fantastic architecture created for expositions, from the Crystal Palace built for the 1851 Great Exhibition and subsequently moved to South London, to the six-hundred-acre site of the World's Columbian Exposition in Chicago in 1893.[29]

Ramsaye's interpretation of the 1895 patent understandably focused on its premonitory aspects. But from the perspective of contemporary media archaeology, we may prefer to see it as a summation of the previous century's range of immersive and illusionistic schemes—drafted by a man poised to seize a timely opportunity to launch the next phase of these practices. For Wells and Paul to become the important twentieth-century figures they were, both had to reinvent themselves in different ways, becoming "realists" rather than the "idealists" portrayed in Ramsaye's beguiling account.[30]

"For the Benefit of Posterity"

Very soon after becoming a pioneer producer, before any of his peers, Paul turned to ponder what value "animated photographs" might have for the future.[31] Surprisingly, amid all his innovative activity, he found time in July 1896 to write to the British Museum, offering a number of "animated photos of London life" as donations.[32] Receiving no reply, he decided to go public and gave an interview to the *Daily Chronicle* in December, which appeared under the promotional headline "For the Benefit of Posterity: Animatographs at the British Museum."

In the interview, Paul argued that films he had recently made, such as *Princess Maud's Wedding* and *The Prince's Derby*, and other "animated photographs of well-known individuals," would be of value to future generations as "an interesting record of current events," like other treasures kept by the museum. He also explained how these combustible objects could be safely stored, in airtight glass tubes encased in metal—

anticipating the issues, and even some of the solutions, that preoccupy present-day film archivists.[33] In January 1897, he wrote again to the museum, asking for "a definite refusal or acceptance" of his offer, and finally received an answer, from Sidney Colvin, keeper of prints and drawings and a close friend of Robert Louis Stevenson. Apparently the museum had been intrigued, even excited, by the proposal, but could not decide where the films would be kept, "since no-one could say to which particular pigeonhole [they] belonged."[34] Following Colvin's gracious acceptance, Paul deposited "an animatographic cylinder, representing the Derby won by the Prince of Wales in 1896 with Persimmon" in February. While references to the project spluttered through the rest of the year in the photographic press, it seems that only this film was ever presented, and much of the comment was skeptical about the "rubbish" that might engulf the museum if more films were to be collected.

No sooner had Paul won this symbolic victory than he turned to another, more pressing aspect of the status issue: how to obtain copyright protection for films? Trying to decide what exactly films were had bemused the museum staff, and the question would remain open for another ten years. If they were considered essentially photographs, then the Fine Art Copyright Act of 1862 would apply. This gave the author of a photograph "sole and exclusive right of copying, reproducing and multiplying" the image. It also allowed that this right could be assigned by the actual maker to a producer or owner. Such transfers had to be recorded at Stationers Hall, where all other kinds of publication were registered. In March 1897, Paul became the first filmmaker in Britain to claim copyright on a film, sending in a frame from the *Sea Cave Near Lisbon* and noting that Henry Short had assigned copyright to him in January.[35]

Later copyright claims by Paul show that such assignments were not always straightforward, especially if there was ill-feeling between the parties, and clarifying the legal status of film would be an interest he pursued into the new century.[36] But the immediate reason for staking a claim must have been Paul's ambitious decision to establish his moving-picture business as a limited company. Advertisements appeared in leading newspapers on 24 April 1897, offering shares in Paul's Animatographe Limited, with the aim of raising £60,000 capital for a stake in what was boldly proclaimed "a new industry."

Scoping the New Industry

The claims made in the prospectus certainly seem inviting. Paul's accountants vouched for a total income of £18,737 from the date of the first entry in his books relating to the Animatograph in March 1896. Although the costs of his other work in instrument making could not be fully separated, an estimated profit of £12,838 on all film activities for the year included "receipts for Exhibitions at Theatres, Music Halls, etc," as well as sales of projectors and films. Assuming capital of only £1,000, this represented a 1,200 percent profit, and there seemed no reason it should not continue, the prospectus claimed, particularly when boosted by two proposed new business activities. The first of these was "the manufacture and sale of animated portraits of individuals," described as a "great and taking novelty, produced at small cost." Studios were envisaged in London and around the country to produce in a "portable and convenient form, records of the natural movements" of clients. The obvious objection to this scheme was that projectors to show such portraits were still rare, but elsewhere the prospectus foresaw wide sales of "an instrument cheaper and simpler than that required for public entertainments" aimed at "amateurs and lanternists."

Paul's other new idea was "animated advertisements." These would be "adaptable to any trade and showing processes of manufacture or use of any commodity in life-like operation." Granted, "stations" to show these would be needed, but once established, "the novelty and attractiveness of such advertisements will probably cause them to supersede all other illuminated devices." The prospectus included an estimate that thirty advertisements, each running five times a week, at a rate of £3 per placement, would bring in revenue of £23,000 a year, yielding £15,000 profit. This, plus £7,000 profit from sales of films and equipment, £7,000 from exhibitions and traveling shows, and £1,000 from animated portraits, brought the summary estimate of profits for the first year of operation to £30,000. Although neither of the new enterprises ever got off the ground, the confidence and vision Paul showed—a mere twelve months after the first commercial exhibition of film—is impressive. Professional portrait cinematography may not have flourished, with amateur filming largely taking over this role, but "animated advertising" has of course far exceeded anything he could have imagined.

One name stands out in the flotation prospectus. Along with a former music-hall director and a solicitor, Nevil Maskelyne of the Egyptian Hall is named as a director of the new company, and one of its assets was to be the rights to a new model of the Animatograph developed by him. This joined four foreign patents that would become assets of the limited company, covering Belgium, Austria, Hungary, and Germany.

One reason for the flotation was frankly admitted: "Mr Paul is no longer able, unaided, to cope with the enterprise in such a manner as to do justice to [its] great possibilities." Undoubtedly, he was stretched to the limit trying to ride this unexpected tide, while also maintaining his existing instrument business. If the new company was formed, he proposed to remain managing director for up to five years, in return for £25,000 in cash and £15,000 worth of shares.

But the "entertainment proprietors, advertisers, photographers and others" toward whom the prospectus was aimed did not respond. Or if they did, it was not in sufficient numbers. Two professional journals cast doubts on the offer—one wondering in a letter how the animated portraits could be viewed by enough people, the other expressing doubts about the future of this whole novelty. The company was never floated, and Paul's business continued as a private concern, so depriving us of any access to its results in the following years.

How should we interpret this failure to attract investment? On 7 May, the future filmmaker Cecil Hepworth, then still a lanternist and a regular contributor to *Amateur Photographer*, admitted that he had been "hopelessly wrong" and took Paul's statement of profits as proof that "animated photography is *not* played out by any means."[37] The historian John Barnes wrote that, "with the benefit of hindsight, Paul's Animatographe Limited offered a very sound investment."[38] But how could conventional investors have known this in early 1897, or indeed have guessed which branch of the "new industry" would prove most profitable? Yet commercial investment in new technologies was not out of the question at this time. We might compare Paul's unsuccessful offer with another in a similar field that succeeded just three years later. William Stanley (born in 1829) had built up a business initially based on drawing instruments, which later capitalized on the craze for stereoscopes, producing a cut-price version in the 1850s, followed by diversification into magic lanterns, surveying and meteorological instruments, and much else based on his numerous patents.[39] By 1891 Stanley had seventeen branches and over 130 workers; in 1900 he raised £125,000 by selling 25,000 shares at £5. No doubt the catalogue of over three thousand tangible, familiar items produced by William Ford Stanley and Co. Ltd, inspired greater confidence than Paul's improbably high profit margin and airy predictions of future growth in a field where few investors could have had any firsthand experience.

Fire! The Threat to Moving Pictures

Within days of Paul's press advertisements setting out a promising future for moving pictures, news arrived of a disaster in France that seemed to threaten the reputation of the new industry—and was long reported in

30. The tragic fire that killed 126 people, many of them titled women, at a charity bazaar in Paris on 4 May 1897 was widely and sensationally reported. Started by careless handling of a film projector's combustible illuminants, it led Paul to introduce a "fireproof" model of his projector and contributed to a growing demand for safety regulation of film shows.

cinema history as if it had done.[40] On 4 May 1897, the cream of Parisian society was gathered for the second day of the annual Bazar de la Charité, raising money for relief of the poor. The event took place in a temporary structure near the Champs-Elysées, on the Rue Jean Goujon, where seventy years earlier Victor Hugo had written *Notre Dame de Paris* (known in English as *The Hunchback of Notre Dame*). Twenty-two stalls decorated in medieval style lined an 80-meter wooden hangar, and by midafternoon the elegant crowd numbered over a thousand.

Moving pictures were a novel attraction at the bazaar, being seen for the first time by most of the aristocrats in attendance. The program consisted of Lumière films, projected onto a canvas screen by a Joly-Normandin machine that used a Moteni lamp burning oxygen and ether, which was stored alongside the projector enclosure. There had been problems the previous day with the lamp, but a successful show was given early in the afternoon. Then, around 4:30 p.m., calamity struck. As the ether container was being changed, a lighted match caused a violent explosion. Within fifteen minutes the entire structure was on fire and over a hundred people died, many of them burned beyond recognition.

The fact that many of the victims had royal connections seemed to add to the sense of tragedy, or at least to the scale of its reporting. The Duchess of Alençon, sister-in-law of the Empress of Austria, was the most widely mourned, with stories told by survivors of her courage in helping others amid the chaos. *Figaro* vividly described the inferno on the following day:

> We saw an unforgettable scene in this huge fire that engulfed the bazaar, where all burned at once, stalls, hangings, floors and walls, men, women, children writhing, the shrieking of the doomed, trying in vain to find a way out, then seared in their turn and falling onto the ever-growing heap of charred corpses.[41]

The most famous image of the fire, published on the front page of *Le Petit Journal* on Sunday 18 May, showed a man helping a woman amid the flames. But the truth was that only six men were listed dead compared with 110 women.[42] A feminist journalist wrote scathingly of twelve men out of two hundred present having behaved honorably, while many of the rest were seen fighting their way past women to escape.[43]

The Paris fire was, of course, widely reported and led to inevitable concern about the safety of moving-picture shows in Britain, as it did around the world. However, it is important to keep this in historical perspective. Theaters where moving pictures formed part of the program were already dangerously prone to fires due to other hazards. One of the first reports of the Charity Bazaar fire in Britain compared it with a fire at the Paris Opera ten years earlier, which had killed two hundred.[44] In the same year, 186 died in a fire at the Theatre Royal in Exeter, while the Iroquois Theatre fire in Chicago killed an almost unbelievable 602 in 1903. Alexandra Palace in North London, near where Paul would build his studio complex, burned down just sixteen days after opening in 1873, but was quickly rebuilt. Such fires were mainly due to the use of oil lamps and other forms of illumination involving combustion, especially limelight, combined with densely packed audiences and inadequate exit arrangements. Their frequency led London's chief fire officer, Sir Eyre Massey Shaw, to publish general recommendations in 1881 — immediately before yet another fire demolished the Alhambra music hall, killing two firemen and nearly injuring the Prince of Wales, who was in the habit of visiting fire scenes. Massey Shaw's report brought about the fireproof safety curtain, separating stage from auditorium, water hydrants, and greater attention to keeping gangways and exits clear.

When the organizers of the Charity Bazaar were brought to trial in August, it became clear that what had caused the fire was one of the projectionists striking a match beside the ether container, rather than the film catching fire. But before this, a Parisian official was quoted as regretting "that the nature of the kinematograph was so little known, and [urging] its being submitted to police regulations."[45] In London in July, the managers of the Agricultural Hall in Islington "insisted that the [cinematograph] apparatus be enclosed in a fireproof chamber," which soon became common practice.[46] Robert Paul responded to the prevailing concern by launching a new model of his projector, now named the Fireproof Animatographe. In addition to its protective casing, which ensured that "only a few inches [of film] can burn ... even if ignited by wilful carelessness" (according to an August advertisement), the projector offered many improvements on previous models. Small, lightweight, with reduced flicker and a new sprocket design that reduced film wear, it took spools on which "a whole series of films may be joined up."

If the fire scare acted as incentive to combine a number of engineering improvements on his earlier equipment, Paul was also prompted to attempt something altogether more ambitious. The "travelling shows" he had predicted in April would use electricity to power the projector and perhaps ancillary lights, and so avoid any need of combustible gas. According to a report in the *Amateur Photographer*, Paul envisaged that the operator "will merely gear the motor of the carriage on to a dynamo which he has brought along with him, and there you are."[47] This was probably a logical path for an electrical engineer to follow, but it may also have signaled Paul's early interest in automobiles: 1896–1897 marked the effective beginning of the British motor industry, with the first Daimler and Humber models appearing. We know that Paul was already a keen motorist by 1906, and it is certainly possible he foresaw the new horseless carriage as a means of further modernizing film exhibition, by means of what the *Amateur Photographer* wittily called "a more than usually animated autocar."

A Royal Flush — Victoria's Diamond Jubilee

One of the future opportunities mentioned in Paul's April prospectus was the "forthcoming Jubilee Fetes to be held throughout the country." Much of Britain and its empire was already preparing for Queen Victoria's Diamond Jubilee, marking her sixty years on the throne. But the theme of the celebration was to be less the queen herself, already feted at her Golden Jubilee in 1887, than the vast empire over which she reigned. Instead of the banquet attended by fifty foreign crowned heads that had celebrated the previous jubilee, the colonial secretary, Joseph Chamberlain, conceived a very public event: a procession around central London

31. Paul with the electrically powered camera he used to film Queen Victoria's Diamond Jubilee (June 1897).

with representatives of all the dominions and colonies in attendance. He can hardly have had filming in mind, but this open-air spectacle would provide an unprecedented opportunity for the growing number of producers and their audiences to participate in an event that had global resonance. Indeed, it was perhaps film that gave it such wide impact. Nearly forty years later, the Jubilee was credited with helping to revive interest in moving pictures, after international film companies' operators lined the route of the procession and subsequently promoted their films.[48]

Paul had acted quickly to secure vantage points to cover the extraordinary procession on 22 June. He recalled:

Large sums were paid for suitable camera positions, several of which were secured for my operators. I myself operated a camera perched on a narrow ledge in the Church yard. Several continental cinematographers came over, and it was related of one that, when the Queen's carriage passed he was under his seat changing film, and of another, hanging on the railway bridge at Ludgate Hill, that he turned his camera until he almost fainted, only to find, on reaching a dark room, that the film had failed to start.[49]

Some of Paul's outlay was recouped by selling off surplus seats he had bought in advance, but the thirteen views of the extraordinary procession that he and his cameramen secured would prove to be best-sellers in Britain and abroad. Following tradition, the procession was led by bluejackets, or marines, pulling field guns, followed by a dozen colonial premiers in carriages, each accompanied by troops in their distinctive uniforms. Squadrons of British and Indian cavalry came next, preceding a group of brilliantly costumed Indian princes, and finally sixteen carriages with members of the royal family, culminating in Victoria in the state landau, flanked by the Prince of Wales and the Duke of Cambridge, with Prince Arthur, who had largely organized the event, bringing up the rear. Altogether some fifty thousand made their way around the circular route, from Buckingham Palace to St. Paul's Cathedral, where a short service was held at the bottom of the steps—in deference to the elderly Victoria, who was unable to climb them—before moving on to the Mansion House, then into South London along Borough Road, then turning across Westminster Bridge to enter Parliament Square and so back to the palace.

The route was densely and colorfully decorated, with large stands erected at vantage points and an estimated three million visitors thronging the city for the occasion. A journalist, G. W. Steevens, summed up what the organizers had made visible in the procession:

> Up they came, more and more, new types, new realms at every couple of yards, an anthropological museum—a living gazetteer of the British Empire. With them came their English officers, whom they obey and follow like children. And you begin to understand, as never before, what the Empire amounts to. Not only that we possess all these remote outlandish places … but also that these people are working, not simply under us, but with us.[50]

Ticket touts, scams that left visitors without places, and pickpockets all flourished. Yet "it was all surprisingly successful," admitted the highly skeptical painter Edward Burne-Jones, who was nonetheless glad when the imperial bombast had finished—"Once in a lifetime's enough for a Jubilee."[51]

Contemporary commentators already foresaw that film would carry this spectacle to wider audiences and into the future. A writer who inter-

32. Queen Victoria's Diamond Jubilee in 1897 was celebrated with a spectacular procession through London that attracted many film companies, including Paul's. A fragment of this part of Paul's coverage has survived in its original red-tinted form. BFI National Archive.

viewed Paul introduced his article in *Cassell's Family Magazine* with an eye to the value of moving pictures as a chronicle:

> This automatic spectator, who is destined to play an important part in life and literature by treasuring up the "fleeting shows" of the world for the delight of thousands in distant countries and in future ages.

Even the showman's paper *The Era* had an eye to posterity:

> Those loyal subjects of her Majesty who did not witness the glorious pageant of the Queen's progress through the streets of London . . . should not miss the opportunity of seeing the wonderful series of pictures at the Empire, giving a complete representation of the Jubilee procession . . . by the invention of the Cinématographe . . . our descendants will be able to learn how the completion of the sixtieth year of Queen Victoria's reign was celebrated.

While Lumière's coverage, led by their star cameraman, Alexandre Promio, included unique material filmed before the actual procession, Paul's films were probably the most comprehensive record of the event as a whole. Throughout Britain, a variety of new moving-picture exhibitors made the Diamond Jubilee the centerpiece of their programs throughout the second half of 1897 and well into the following year.[52] Producers and distributors experienced a boom in sales, and resorted to renting rather than selling outright the most popular items — possibly for the first time — to secure maximum returns while the event was still popular. Outside Britain, across the empire, which also provided a new market for films from the "mother country," it would prove even more popular.[53]

As filmed images of Victoria's Jubilee traveled around the world, followed four years later by images of her funeral, many among Paul's community of technologists were already contemplating how wireless telegraphy would make possible instantaneous communication over vast distances. In 1901 his former teacher and patron, the physicist William Ayrton, chaired a meeting addressed by the radio pioneer Guglielmo Marconi at the Society of Arts, and described the prospect of calling "from pole to pole" as "almost like dreamland and ghostland," although imminent.[54] Meanwhile, film already enabled audiences to participate by proxy in distant events, and thus served as a time- and space-shrinking machine. Louis Lumière may have believed his invention "had no commercial future," but Paul saw from the outset that the records of contemporary life it provided would become increasingly valuable, a belief that underlay both his approach to the British Museum and the attempted flotation of a company to develop wider uses of film. From our perspective, of course he was right. The surviving portions of his and others' Diamond Jubilee coverage are what connect us with those original spectators of the great 1897 procession.

CHAPTER 5

"True Till Death!"

FAMILY BUSINESS

Amid all the hectic activity of 1897, Robert Paul found time to marry; and as with his choice of career, the decision owed nothing to his family. His bride was a dancer from the Alhambra company, Ellen Daws, who had appeared under her stage name "Ellen Dawn" in *The Soldier's Courtship* eighteen months earlier. In that film, she played, not the flirtatious object of Fred Storey's attentions, but the fat busybody who interrupts them and is duly tipped off the bench. Apart from this disguised appearance and several other rumored "character" roles in Paul's films, there is no known photograph of Ellen.[1] Irene Codd undoubtedly knew the Pauls, and was closer to Ellen, so is a valuable (if uncorroborated) source of information about what led up to and followed the marriage on 3 August.[2]

The daughter of a master cabinetmaker, Augustus Daws, Ellen (or "the redoubtable Nellie," as Codd calls her) had, as a child, told her parents that she wanted to go on stage, and when they rejected the idea she turned to an aunt, Mrs. Nye Chart, for help. Chart had become the first female theater proprietor in Britain after the death of her husband, John Chart, who had renovated the Theatre Royal in Brighton.[3] Mrs. Chart apparently paid for Ellen's training, which led to a position at the Alhambra, where she came into contact with "the mad-looking man with the silk hat jammed on the side of his head, and his coat-tails flying in the breeze."

It would be poetically satisfying to imagine the pair met during the making of *The Soldier's Courtship*. Or even that Paul's courtship was in some way reflected in the autumn 1897 film *Cupid at the Washtub*, in which a flirtatious groom tries to steal a kiss from a servant girl, and she dunks his face in the washtub.[4] Presumably a liaison was suspected, or expected: Codd records that Ellen was chaperoned nightly from the theater by her mother. But before her mother died early in 1897, she apparently confided that she had dreamed of her daughter being married to Paul. There were invitations to supper with him, which Ellen declined, and the

33. The Alhambra Theatre ballets were famously lavish. Ellen Daws was a member of the company when Paul started his daily shows in March 1896, and appeared in *The Soldier's Courtship* in April.

gift of a gold bracelet "with a design of lovebirds." Then Paul asked her to repeat the words "True till death!" immediately before announcing to a party given by the theater's music director Georges Jacobi that they were engaged.

Codd claims that Paul's family was unimpressed by his choice of an actress, which would not be surprising in light of the then still dubious reputation of the profession. Ellen was apparently "looking after an invalid uncle" when Paul invited her for a drive, which led to St. Giles' church, where his solicitor was waiting to act as witness at their marriage.[5] An equally unannounced wedding luncheon was held at Frascati's, an opulent restaurant on Oxford Street,[6] before the couple drove to Paul's upper-floor flat in High Holborn, which he had partly fitted out, leaving "the drawing room ... empty so that Mrs Paul could furnish it herself with whatever furniture she liked." Codd's account of the marriage continues with an anecdote about Ellen's seeing a hansom cab regularly parked across the road, which turned out to be Paul's mother keeping a discreet eye on her son. Before long she is invited in and shares their lunch, and the ensuing rapport leads to regular visits to the parents, now "living in a mansion at Loughton, Essex," which is indeed where Paul's parents and siblings were living in 1901, according to the census record.

Codd frames her memoir of Ellen by asserting somewhat unexpectedly that she was "the business brain" of the partnership, who "made [Paul] the success financially that he became." She writes that he "invested a large sum of money in [Ellen's] name," on the basis that she had "earned a great deal for him" and had managed his instrument-making business while he visited America to open a branch.[7] However improbable this may seem, in view of Ellen's lack of previous technical or commercial experience, a professional colleague, W. H. Eccles, noted in Paul's obituary that Ellen had played a major part in running the film business: "His wife was producer, stage manager or principal lady in many a playlet for which her expert knowledge eminently fitted her."[8] Failing to credit Ellen with a role in Robert's hectic production career fits with the tradition of seeing early cinema as an exclusively male industry. Women's work at this time was largely unacknowledged, or viewed with disfavor. It seems highly probable that Ellen's early debt to Nye Chart had opened her eyes to the potential for women to manage, which she seems to have done. But with the exception of Alice Guy—now widely regarded as Leon Gaumont's first "director"—and a very few others, women have received little or no recognition for the contribution they undoubtedly made to much early film activity.[9] Yet, in a rare case where detailed documentation has survived, Patricia Cook has been able to show that William Slade, one of Gaumont's earliest customers in Britain, depended on his daughter Mary to staff and probably organize his wide-ranging tours during 1897–1898.[10]

Just under seven months after the Pauls' marriage, on 22 January 1898, their first child was born and named Robert Newton Paul. The unusual second name suggests an homage to the founder of modern physics by a young science graduate; Paul would also give the name to a road near his studio, probably as a memorial.[11] Tragically, their son's life was cut short at just seven months, on 23 August, with the cause of death recorded as "diarrhoea, convulsions," and his father "present at death." Years later, the story gained currency that the infant's death had been linked with an accident during filmmaking—and had been responsible for Paul's abruptly abandoning the film business.[12] Codd's memoir confirms this link in emotive terms:

> [Ellen] cursed the film business, as it was responsible she always said for the tragedy of their lives. Paul had made up a bottle of fake milk in a baby's bottle for one of his comedy films, and a nurse-maid had given it to the baby boy and poisoned him.

Their son's death did not put an end to Paul's career in film, most of which was still to come after August 1898. However, it is possible that Codd's account can be linked with one of the "paired" subjects that appeared along with the elaborate multiscene films of that summer, and marked Paul's major innovation as a producer. *The Sailor's Departure* and *The Sailor's Return* form a pair that could be shown separately or consecutively.[13] In the first, a sailor says farewell to his wife; in the second he arrives home to see an infant born during his absence, which may well have required a feeding bottle as a prop. The Pauls were exceptionally active in front of as well as behind the camera during the summer of 1898.[14] And because the film had such a terrible indirect impact on its makers' lives, it may have been the last in this exceptional burst of production, followed by over a year without any fictional films being issued.[15] Apart, that is, from a solitary Christmas subject released in November, *Santa Claus and the Children,* which would indeed have been a poignant subject to produce at this time.

In her manuscript, Codd refers to two further children born to the Pauls, who turn out to have been Lizzie in 1902 and George Rollason, given Ellen's mother's maiden name, in 1903. Both, however, died almost immediately due to birth complications.[16] To have suffered three such losses must have been devastating, although it does not appear to have diminished Paul's commitment to his second demanding business— perhaps nonstop activity distracted husband and wife from their grief. Codd describes the Pauls as a couple "who would have made excellent parents," and whose "affection was largely given to Paul's nephew, whom they brought up as their son."[17]

Paul's will included provision for two nephews, George Ivens Paul and David Lyon Paul.[18] George, the son of Robert's brother Arthur, was born in 1911.[19] Arthur died in 1922 in Coventry, but whether the boy lived with Robert and Ellen from an early age is unclear. Robert's younger brother, George Herbert, had died in 1919, apparently unmarried and with "Arthur in attendance." Both brothers are recorded in the 1901 census as working in shipping, Arthur as a "ships stores dealer" and George as a "ship-owner's clerk," which suggests that both had remained close to their father, presumably returning to work with him after their South African war service. Their father, George Butler Paul, is listed in Lloyds Register as the co-owner of up to six merchant ships, but only between 1897 and 1907, which does not support the claim by his son's obituarist that "his father was a London shipowner, whose ships sailed out of the Pool of London to the Baltic and the Levant," with Robert taking "long trips during vacations on his father's ships." Assuming this was not invented, he was presumably able to take such voyages because his father already worked in shipping, even if he was not yet an owner.

Travel certainly came to play an important part in Paul's life. Initially around Britain, during the early years of film screening; then to Sweden and Norway, where he would return much later in the 1930s. And during the years of developing the electrical instrument business, he visited New York on at least one occasion, in 1913. What is missing from the image that has survived is any sense of Paul's personality, or indeed the motivation for his years of intense effort between 1895 and 1914. That Ellen worked closely alongside him, helping to manage both businesses, seems highly plausible. He would play a leading part in the Kinematograph Manufacturers Association after 1906, deputizing for its elected chair, which was presumably when he acquired the nickname "Daddy Paul." A cartoon published in a trade journal in 1905 shows him amid fellow producers, at the center of heated discussion over how to respond to price-cutting by Pathé;[20] he appears too in a group photograph at the 1908 negotiations in Paris, which led to several producers, including Paul, leaving the business (see figures 65 and 66).

The few authenticated photographs we have of Paul show him as serious, even solemn. Eccles ended his obituary with the frank observation that "his reserve and shyness and a kind of talent for fading away unnoticed were exceptional."[21] Yet the same obituary records that when the 1896 Derby film was shown at the Alhambra, "Paul was called before the curtain many times and received a great ovation." It also affirms, clearly from personal reminiscence, that Paul played an active part in the Muswell Hill Studio, where he "painted all the scenery at night, 'after the day's work was over' as he said." But apparently not every night. Codd paints a vivid picture of the Pauls enjoying regular visits to one of London's West End institutions, the Café Royal, near Piccadilly Circus. Founded in the 1860s by a refugee French wine merchant, this multilevel restaurant would become, she rightly observes, "the haunt of Bohemian London," with a celebrated list of patrons that ranged from royalty to most of England's literary celebrities, including such controversial figures as Oscar Wilde and Aleister Crowley.[22] Max Beerbohm vividly recalled its aura:

> a haunt of intellect and daring, the domino room of the Cafe Royal. There, on that October evening—there, in that exuberant vista of gilding and crimson velvet set amid all those opposing mirrors and upholding caryatids, with fumes of tobacco ever rising to the painted and pagan ceiling, and with the hum of presumably cynical conversation broken into

34. *The Café Royal.* Painting by William Orpen, 1912. Ellen and Robert Paul fre-
quented this "haunt of Bohemian London," according to Irene Codd, especially on
occasions such as New Year's Eve and when being entertained by their young theater-
going friends.

so sharply now and again by the clatter of dominoes shuffled on marble
tables, I drew a deep breath and, "This indeed," said I to myself, "is life"!

The Pauls were apparently often entertained there by "their many young
friends" before frequent theater visits, which suggests a more sociable
aspect than Eccles describes. Tantalizingly, Codd claims that "they were
the leading lights at the Café on New Year's Eve," which should no doubt
be connected with the many examples of boisterous and comic behavior
in Paul's films.

But perhaps the most remarkable evidence of the Pauls' sociability
comes from a trade journal in 1906. The account of a summer "works outing" for the studio staff, clearly written by a participant, is worth quoting
in full:

Mr Robert W. Paul entertained his numerous staff on the 21st at his country retreat. The party was taken down in large brakes, and after a look
around the estate and lovely gardens, sat down to an open-air repast.
After the meal, the visitors were photographed, and an adjournment was
made to the cricket field, where a hearty game was indulged in. Others
went in for quoits, fishing and boating, until the failing light brought the
whole party once more together, when an al fresco concert was begun,
with piano, violin &c. The return to dreary London had to be commenced

35. *Come Along, Do!* (1898). Based on a popular theatrical sketch, Paul's first two-scene film shows a country couple outside an art exhibition, before they enter and the man's attraction to a nude statue leads his wife to scold him. An unsourced note in BFI records suggested the parts were played by Robert and Ellen Paul, although the man appears unlike known portraits of Paul.

all too early, but everyone will remember the most enjoyable outing, to which Mrs Paul's kindly attention largely contributed.[23]

This day in the country certainly suggests a generous and sporting host, and confirms how much Ellen continued to be involved in the life of the film studio. Many of the six "members of various departments" mentioned in the report would go on to play important roles in the industry after leaving Paul's employment, suggesting the company served as a kind of training ground for future British filmmaking.[24] Codd evokes the life of the studio:

> How the little company must have enjoyed themselves, and how those of them that are still alive today must smile when they look back and remember the little shed with the glass roof, and Mrs Paul with her sewing machine.[25]

If the information about his wife's appearances in Paul's films is correct, the two that survive after *The Soldier's Courtship* both show couples in relationships where Ellen is firmly in charge. In *Come Along, Do!*, it is the wife who interrupts her husband's inspection of a nude sculpture, and when a burglar enters the bedroom in *His Brave Defender*, she boldly tackles the intruder, while her husband hides under the bed. Without reading too much into these — or being able to view *The Lover and the Madman* — Ellen's roles fit well with the accounts given by Codd and Eccles of her importance to the filmmaking process.

Codd describes as well, perhaps with some dramatic license, the impact of the studio on the surrounding community of Muswell Hill, where news of a film about to be shot in the streets "gave great joy to the local children":

> Mothers complained that their children were playing truant from school with alarming frequency.
>
> "Paul's going to make a picture, I'm not going to school!" "Me neither" was the cry.
>
> The parson and the schoolmaster called at the works, and asked Paul to keep it quiet when he was going to make a film.
>
> "Keep it quiet," said Mr Paul. "Why they've got a bush telegraph system. They know when I'm going to make a film long before I know it myself." . . .
>
> There was a permanent spy outside the gates of the studio equipped with roller skates, so that when Paul emerged from a cab, the clattering Mercury rushed off, the word was given, and every child for miles around was on Paul's track.[26]

The Miller and the Sweep, Codd reports, was filmed "outside a cottage at Muswell Hill," and another subject, "taken at Highgate station of police chasing a convict[,] was so realistic that the public joined in, rendering the film useless as the crowd was too great." This could refer to *The Fatal Hand*, in which the crowd that joins the chase after an escaping killer is indeed large, although the film was certainly released.[27]

Codd was too young to have witnessed such scenes herself, so must have heard reminiscences, most likely from Ellen. The stories provide some sense of how Paul's filmmaking was woven into his family life, especially after he and Ellen moved to Muswell Hill. We know that they lived in rooms on High Holborn at least until 1901, in Muswell Hill by 1903, and until 1908 in various houses on Colney Hatch Lane, close to the Sydney Road studio and factory site. But it was not until Paul had given up the film business that he built 51 Sydney Road, to a custom design.[28] Attempts to locate the "country retreat" have so far failed, although the report of the convivial company outing fits well with Codd's firsthand account of the Pauls' social and recreational life. Apparently they owned "a motor launch, which was kept at Maidenhead," and they "frequently spent weekends chugging up the Thames."[29] Again, this corresponds with

36. While Paul was filming the launch of the battleship HMS *Albion* on 21 June 1898, his electrically powered camera panned across the scene and caught what appears to be a close-up of his wife, Ellen. (See also caption 35.)

the number of films involving small boats, including the very early res-cue drama *Up the River*, in which a man—rumored to be Paul himself— jumps in to rescue a baby that has fallen from a steam launch. It would also shed light on the series of films taken from a launch further down the Thames, which included the rescue of victims from the *Albion* disas-ter in 1898.[30]

Despite abandoning the film business, the Pauls seem to have decided to put down roots in North London, first submitting plans to the local authority to build a detached house on Sydney Road in 1911. The house at number 51, initially known as Newton, was finished in 1912, and included a number of distinctive design features, including window catches appar-ently styled after the "unipivot" bearing at the heart of Paul's innovative galvanometer. At some point after 1918, however, the Pauls moved to a very grand house in Addison Road, Kensington, close to where the re-search chemist and future president of Israel, Chaim Weizmann, lived during the war and where David Lloyd George would come to live in 1928. By this time, Paul presumably had no day-to-day responsibility for the factory that continued to operate in Muswell Hill. Moving from the relatively modest Sydney Road house also allowed the couple to have a larger staff, and gave Paul space for a workshop in the basement. An em-ployee who was deputed to work as his assistant in the mid-1930s recalled

37. At some point after 1918, the Pauls moved from Muswell Hill to a mansion on Addison Road in West London, near the scene of his early exhibition experiences at Olympia. Photo: Ian Christie.

"six or seven servants," two of whom would bring the men tea.[31] Mrs. Paul was described as "a very pleasant, large lady, liked by everyone." Apparently, "she had started a sponsored clinic not far from the factory," which may have been prompted by the couple's unfortunate experience of infant mortality.

Paul remained a keen motorist, as he had been since the very early years of automobiles, and Codd tells how their "small Rolls Royce ... was nearly the cause of Mr Paul's death."[32] Working on the car with the garage doors closed, "the carbon monoxide fumes overcame him, and he fell to the ground unconscious." Realizing he had been missing for some time, Ellen found him on the ground and "dragged him into the open air by his ankles," but his recovery was slow, and the accident "affected his heart." One of the last public sightings of Paul had a motoring context, when he was recognized unloading his car on the dockside in Norway. Motoring and travel evidently remained lifelong enthusiasms.

Codd writes cryptically that Paul had more contact with Queen Victoria and King Edward VII than has been recorded, referring to invitations to show films at both Windsor and Buckingham Palaces. Her claim that "he refused a knighthood" was echoed in an obituary, although without any indication of when this might have taken place.[33] If not for his film activities—knighthoods have been rare for British filmmakers—it could plausibly have been in recognition of his war work or his role in electrical science. Horace Darwin, founder of the Cambridge Scientific Instrument Company, with which Paul's firm merged in 1918, was knighted in that year, largely for his contribution to wartime technology.

Paul never lost contact with the scientific world, despite the heavy demands of the film business. Playing various roles in the Royal Institution brought him into contact with some of the leading scientists of the day, notably its charismatic president, Sir William Bragg, whom he had probably met during wartime research activity, and Bragg's son Lawrence, who shared the Nobel Prize for physics with his father in 1915. He was even able to combine these interests, in a series of scientific demonstration films he made for leading scientists.[34] However, there is one further intriguing, and so far unexplained, allusion to what may be another, hitherto unknown dimension of Paul's life. When he gave his paper to the British Kinematograph Society in 1936, marking the climax of a period of reminiscence about his film career, the proposer of a vote of thanks, Captain Paul Kimberley, spoke of having known Paul

> in the days which he termed "Reformation." He was filming Biblical stories—"the Wise and Foolish Virgins," "The Prodigal Son" etc—and I remember I had to show them to the Bishop of London and other Church dignitaries as examples of what could possibly be done with Biblical subjects.[35]

Paul's known productions do include *The Prodigal Son* (1908), although no interpretation of "the Wise and Foolish Virgins." An early catalogue advertised *Buy Your Own Cherries* (1904) as "a splendid illustration of the famous temperance story, Buy Your Own Cherries," describing it as "specially suitable for sacred entertainments." But otherwise, this knowing remark by a well-connected industry figure, then managing director of the National Screen Service, remains puzzling. Why would Paul be

38. Robert Paul's gravestone in Putney Vale Cemetery, South London. The inset bronze motif (*top right*) is of the "unipivot" mount in his most successful electrical instrument range, which is also echoed in the design of window catches (*bottom*) for the house he had built at 51 Sydney Road in 1911. Photos: Ian Christie.

exploring "what could be done with Biblical subjects," and what interest could the then bishop of London, Arthur Winnington-Ingram, have had in these films? The answer might be that, as a senior and respected figure in the industry, Paul was trying to counter the bishop's shrill campaign against "immorality" on stage and screen, and specifically against cinemas as places for gay sexual encounters.[36]

Just seven years after his paper to the BKS, Paul would be dead at the age of seventy-three, having spent his last months in a nursing home in South London. He was buried in Putney Vale cemetery, with an unusual headstone, featuring a motif based on the design of his most successful electrical instrument, the unipivot galvanometer. It also carries the phrase "True till death," which thanks to Irene Codd we know had been a motto for Robert and Ellen since 1897. Ellen would live on for another eleven years, not dying until 1954, although initially troubled by an unexpected feature of her late husband's will.

Paul's estate was valued at £215,456, from which he specified small bequests to two sisters, two nieces, two nephews, and several former employees. The balance of his estate was to establish the R. W. Paul Instrument Fund, which would support awards for innovative engineering projects. Ellen received mainly property and effects, without any specified sum of money, and a year later she petitioned the High Court successfully for a reallocation, on the grounds that the will did "not make reasonable provision" for her maintenance. The result was an agreed annuity of £1,800. Why Ellen had to go to such lengths to secure an adequate income remains puzzling. Had Paul assumed that the property left to her would suffice? It is hard to believe he intended to leave her in relative poverty. Meanwhile, the Paul Instrument Fund, now administered by the Royal Society, continues to disburse awards.

CHAPTER 6

Home and Away

NETWORKS OF NONFICTION

The explosion of interest in moving pictures during 1896–1897 was not confined to Britain, and Paul's early position as the largest open-market supplier of both projectors and films, based in the world's greatest trading center, offered unusual opportunities for expanding his business internationally. He would later recall "customers ... from nearly every country" coming to his office, where "each insisted on waiting until his projector was finished."[1] Paul specifically mentioned callers from Turkey, Spain, and Denmark, mostly in the context of anecdotes about the mishaps that befell early customers:

> Four Turks, speaking little English, came daily for weeks.... Finally they found that the attractiveness of the night life in London had led to the complete exhaustion of their financial resources. A gentleman from Spain, anxious to return quickly ... proved too impatient to learn how to centre the arc light, and left with his projector, unboxed, in a cab. Arriving in Barcelona, his first attempt at projection failed, so the disappointed audience threw knives at the screen and wrecked the theatre. He himself retired to serve a term in a Spanish prison.[2]

Forty years later, such stories evoked the raffish pioneering days of film shows, with disasters recalled more often than successes. In reality, Paul's projectors and films were probably toured at least as widely and successfully as the Lumière Cinématographe during the first year of commercial film exhibition.

The basis of these shows was nonfiction "actualities," as it would remain for most of the following ten years, despite the steadily increasing proportion of anecdotal fiction films available. A program of "animated photographs" in the autumn of 1896 would consist of around twenty individual subjects lasting approximately the same number of minutes, as exemplified in the flyer for Paul's program at the Alhambra (see figure 26).[3] By 1900, when he launched the series *Army Life*, the running time of this—admittedly exceptional—program was 75 minutes. Until the introduction of "long films" around 1910–1911, bitterly contested by many in the film trade who insisted that audiences "wanted" a varied program (which was also cheaper than the new, largely imported "features"), normal showings probably remained around 70 to 80 minutes.[4] A program might include some novelty items, such as films accompanied by live or recorded songs, and pictorial slides commonly appeared between films, especially as, until the arrival of the multireel feature, most venues had only one projector, often with a slide carrier alongside the film mechanism, as in models Paul offered until at least 1906.

The cultural climate of the Victorian era included a wide range of instructional and informative presentations, many taking the form of a "visit" or a "journey," often narrated by a presenter who was usually known as a "lecturer" (which did not then have a purely didactic connotation). One of the most famous of these in Britain was Albert Smith, who presented his "Ascent of Mont Blanc" two thousand times at the Egyptian Hall in London, over seven seasons in the 1850s. Smith, who had actually climbed Mont Blanc, drew on and elaborated his memories of the feat, with a series of moving panoramas as illustration, rolling horizontally for the journey to the mountain and vertically for the ascent.[5] A famous *Illustrated London News* picture of Smith performing in 1852 probably conveys what a Victorian travel "lecture" was like, long before the genre began to incorporate moving pictures. When it did, established international lecturers, such as the American Burton Holmes, were quick to take advantage of the new asset.[6] Early film producers followed their lead in producing film "tour" series, with Paul and the Lumières among the first to do.[7]

What was crucial to these shows for at least a decade was the role of the showman, who would select, combine, and often link the short films that made up the program. Charles Musser and Vanessa Toulmin have written about some of the outstanding showmen, such as Lyman Howe in the United States and A. D. "Edison" Thomas in Britain.[8] "Howe's organisation of [the] films relied on sophisticated editorial strategies," Musser writes, while evidence from Thomas's hyperactive advertising of his shows points up his constant fine-tuning of contents for specific audiences. Paul's catalogues contain a running thread of advice and support for those compiling programs from his films, ranging from the title slides, "artistic" and plain, offered in 1901, to the apparently pioneering inclusion around this time of titles printed as part of the films themselves. For the

Army Life series, Paul recommended title slides "in cases where a lecturer is not desired," which suggests spoken accompaniment was by no means universal.

In this earliest period, 1896–1897, when "animated photographs" were still a novelty, the presence and personality of showmen was often a vital element. Most of those presenting film shows were already involved in other forms of entertainment, and brought that experience to bear in making the most of the new medium. Probably the most numerous were lanternists, who early on started to incorporate moving pictures in their shows, with many eventually becoming exclusively film exhibitors. But there were also conjurors and magicians, such as David Devant and Carl Hertz, both among Paul's first customers. And of course once Méliès had his equipment from Paul, he started introducing his moving pictures into the shows at his Théâtre Robert-Houdin in Paris.

Touring the Modern Mystery

Carl Hertz was the stage name of Louis Morgenstein, born in San Francisco in 1859, who would later style himself "the Modern Mystery Merchant." As noted in chapter 2, Hertz was the first to take a Paul projector on an extensive journey, which would eventually include South Africa, Australia, Ceylon, India, China, Japan, the Fiji Islands, and Hawaii. Thirty years later, he gave his colorful account of "persuading" Paul to part with one of his projectors. Once embarked for his first port of call in South Africa, Hertz discovered he had paid too little attention to Paul's instructions for operating the projector. It was only with the help of a ship's mechanic that he succeeded in presenting what would be the first-ever film program at sea, as a novel addition to his conjuring act.

Arriving at Cape Town, Hertz traveled to Johannesburg, where he opened at the Empire Theatre of Varieties on 20 April 1896, although he did not present moving pictures until nearly three weeks later, at a press show on 9 May. It would seem he was still having problems with the unfamiliar equipment, despite claiming airily that he had "got the rights for the cinematograph for South Africa and Australia from Edison and Robert Paul," although adding truthfully that "the machine I have is exactly the same as the one in use at the Alhambra." Two days later, Hertz began a month-long run, presenting as part of his act a selection of Paul subjects, including *Highland Dances*, *Street Scenes in London*, a *Military Parade*, a *Trilby Burlesque* from the Alhambra production, and *The Soldier's Courtship*. Since the last two had not been filmed before his departure, they at least must have been sent later by Paul—making an-

39. Newspaper advertisement for Carl Hertz's debut with the Theatrograph at Sydney Tivoli, Autumn 1896.

other "first" in long-distance film transport. Hertz then went on to visit Pretoria, Bloemfontein, Kimberley, Durban, Maritzburg, East London, Kingwilliamstown, Grahamstown, Port Elizabeth, and Cape Town in a densely booked tour.[9] En route he managed to acquire some twenty Edison Kinetoscope films, which he could also run on the Animatograph.

On 20 July, Hertz set sail for Australia and disembarked in Tasmania, reaching Melbourne on 11 August. This was toward the end of the "marvellous Melbourne" era, when the city was at the peak of its social and economic success. On his second visit to Australia, Hertz was booked to open at Harry Rickards's Opera House on Saturday, 15 August, although he did not immediately reveal the new attraction. But at a special midnight session on Monday, after the normal evening performance, Melbourne's theater world was treated to the country's first screening. The *Melbourne Herald* described the event evocatively on the following day:

It was after the opera was over. The occupants of the galleries had clattered noisily downstairs, the pit and stalls had cleared, demurely, and the crutch and toothpick dress circle had quitted, talking of oysters at Madame's. . . .

And then, as if by magic as wonderful as any of Mr Carl Hertz's clever illusions, stalls and dress circle filled up again. . . . The audience, there by special invitation to witness Mr Carl Hertz's new illusion, the Cinematographe, was a highly critical one. It consisted of all the members of Mr Rickards' Melbourne company, and of many other clever artists now in the city.... And they did not spare Mr Rickards, Mr Hertz or the leader of the orchestra....

The ringing up of the curtain, however, showed that actors can teach a useful lesson to certain classes of theatre goers. The audience became quite silent and attentive, breaking out into loud applause as Mr Hertz brought out pretty and laughable and wonderfully interesting scenes.

In principle, it is the kinetoscope of Mr Edison. In practice, it is a marvellous improvement upon it. Life-size figures and pictures, true to nature, are shown upon the canvas. A couple of scenes from a comical Trilby called forth much laughter, and scenes of London streets and bridges, with crowds of traffic, omnibuses moving rapidly, hansom cabs dashing speedily, fairly brought down the English part of the house.

And as the orchestra played "A Home on the Ocean Wave" the audience burst forth into general applause, and gave Mr Hertz a special call.

As this and other accounts noted, the Kinetoscope was already familiar to Australians, having been shown and widely toured for over a year. Projection, however, was recognized as a "marvellous improvement," and widely reprinted accounts of the midnight preview predicting that it would "create a sensation" proved accurate. When Hertz added the "cinematographe" to his public performance on the following Saturday, it became the talk of the city. On 4 September, the *Sun* reported that "no alteration of bill has been found necessary at the opera house, for as regards crowds, the cry is 'still they come' to marvel over the cinematographe." The season was extended until 17 September, after which Hertz and his wife traveled to Sydney, where they opened at another of Rickards's theaters, the Tivoli on Pitt Street, on Saturday, 19 September.

By now Hertz had a standard press release, sent to newspapers ahead of his appearances, which gave a mixture of factual and fanciful information about the "cinematographe," as well as descriptions of the main subjects on offer.[10] Billed as "a combination of photography, electricity and stereopticon," it was described, not inaccurately, as "an invention partly of Edison's, and partly of a man named Paul," whom Hertz located at "the Royal Institute in London," before boasting that he had "added some improvements of my own." Invoking "electricity" seems to have been entirely opportunistic, since neither Paul's camera nor his projector was electrically powered at this time. Edison's Kinetoscope had been battery-driven, but Hertz certainly would not have been able to count on a supply of batteries on his extensive travels. Hertz stressed the attraction of busy streets in London, "where you can see the 'buses, cabs &c moving along as though gazing at the actual scene," along with key events in the English calendar—the Derby and the Oxford and Cambridge boat race—that would have been well known to many in colonial Australia. Hertz's account of the actual films is typical of many in this early period, emphasizing the vast quantity of individual frames involved (1,400, "passing through the lens at about 600 a minute") and claiming the cost to be £30.[11]

In Sydney, however, Hertz would not enjoy the exclusivity he had in Melbourne. For earlier that month, Joseph McMahon had held a private demonstration of what was described as a "kinetomatographe," and on 28 September the official Lumière operator Marius Sestier opening the "Salon Lumière" with his Australian partner Walter Barnett, laying much emphasis in the advertising on the projector's being an "authentic Cinematographe . . . direct from the Lumière manufactury."[12] No direct comparison between Paul's and the Lumières' equipment has been found in the contemporary record, but some reviews emphasized the pictorial quality of the latter, and it has been suggested that because the "Salon" was *not* a theater, it attracted "a more elite patronage, which soon included the Governor and various church dignitaries."[13] While Sestier remained in Sydney and began to film local subjects, as was the roving Lumière operators' practice, Hertz finished on 14 October and set off on a tour around Queensland, not showing film at every show, before returning to Melbourne and Sydney at the end of the year and setting sail for New Zealand in mid-February 1897.

Once again, this was not virgin territory. Several New Zealand cities had already witnessed film screenings, including Auckland, where self-styled "Professors" Hausman and Gow had shown a program in October 1896 at the same opera house where Hertz started a week's run on 20 February. The earlier show was billed as by "Edison's Kinemotagraph" and formed part of a vaudeville program. New Zealand historian Clive Sowry has surmised that the films shown were a mixture of Edison Kinetoscope subjects (including *The Milk White Flag* and *Skirt Dancer*) "and English films (possibly produced by R. W. Paul) obtained by Edison's agents."[14] After Auckland, Hertz would go on to play twenty-seven venues in New Zealand over the next four months, including three in Christchurch and a return to Auckland in May. The stamina required to cover this much ground seems remarkable, as does the fact that a population of just over seven hundred thousand apparently sustained at least ninety individual performances.

Hertz left New Zealand for Tasmania at the end of June 1897, and launched into another intensive tour that would take him from Hobart to Melbourne, then on to Adelaide and Perth and a further twenty-seven venues across Australia, before sailing to Ceylon on 18 December. From here, he continued to India, China, Japan, Fiji, and Hawaii. The whole venture appears to have been highly profitable, with shows in Australia and New Zealand yielding £10,000, according to Hertz's 1924 memoir. But even if he did not show films at every performance, it is hard to believe the original prints were in viewable condition after such intensive use.

Spain, Portugal, and Egypt

If Hertz made the most intensive and extensive use of his early Animatograph and Paul's first films, there were other screenings in many different countries, often coinciding with Lumière Cinématographe exhibitions. First off the mark in continental Europe was Edwin Rousby, a Hungarian-born showman who had been performing in France and Spain with his "electric orchestra" before he encountered Paul's projector while touring music halls in Britain in March 1896.[15] Recent research on this shadowy figure indicates that he and his wife had anglicized their names while appearing at the Folies Bergère in Paris, becoming Professor Edwin and Maud Rousby, and that they abruptly changed their act to present Paul's Animatograph in Madrid.[16] The venue chosen by Rousby was a theater known as the Circo Price, after its English founder, or Circo Parrish, after the long-serving manager, and here the Animatograph show opened in early May to popular acclaim. According to a press report on 18 May, "For the tenth *soirée fashionable* all seats were taken."[17] Paul's colorful account of the Spanish debacle (quoted earlier) probably referred to an attempt by one of Rousby's assistants, Alexandre Bon, to take a second projector to Barcelona for a show at the Teatro Eldorado, which indeed failed, no doubt incurring the audience's displeasure.[18]

Rousby was apparently due to continue as part of a mixed company presenting equestrian and mime acts in other Spanish cities (Barcelona, San Sebastian, Vittoria), but on 12 June he contracted to appear at the Real Coliseu in Lisbon, opening on 18 June and serving as an entr'acte to the main presentation, a popular three-act operetta. Projection was from behind a moistened screen, as was common practice in large theaters at this time, to maximize the brightness of the image by shortening the projector's "throw."

The shows were successful enough for Rousby to publish a note of thanks to the public, and to go on tour in July and August to Porto, Espinho, and Figueira da Foz. But early exhibition could be fiercely competitive, and while he was away from Lisbon, another theater decided to buy its own projector and offer a rival show. Their publicity attacked Rousby, accusing him of using a fake projector, even when he demonstrated an "Animatograph colossal," giving larger images. For the second two weeks of August, the shows were in competition, until on 30 August the Amelia Theatre caught fire, putting an end to its program, and allowing Rousby to continue uncontested until mid-January. He appears to have been receiving a regular supply of new films from Paul in London, including, at the end of August, the spectacular series *Tour in Spain and Portugal*, taken by Henry Short at Paul's initiative.

Short had taken on a role similar to that of the Lumières' traveling operators, who filmed at their own initiative and showed the results in the places they visited. Short's eighteen films of scenery and customs amounted to an outline gazetteer of the Iberian peninsula, guided by the fact that travel by sea along its coast was easier at the time than travel by road or train.[19] The series included city scenes taken in Madrid, Cadiz and Seville, and Lisbon; two films of a bullfight; two exotic dances (the Fado and an Andalusian dance); and a Portuguese train "bringing bathers to the coast." The great success of the series, however, was *Sea Cave Near Lisbon*, as already noted. By 1901 this had achieved classic status, with an eloquent catalogue description:

> This famous film has never been equalled as a portrayal of fine wave effects. It is taken from the interior of a great cave, looking over the ocean. Big waves break into the mouth of the cave and rush towards the spectator with the finest and most enthralling effect.[20]

Waves breaking on the seashore had already made a strong impression on early spectators. Acres and Paul's *Rough Sea at Dover* (1895) had been the undoubted hit of the first Edison screening on Broadway in April 1896, inspiring one of the most eloquent published responses to any early film:

> [A roller] came down the stage, apparently, increasing in volume, and throwing up little jets of snow-white foam, rolling faster and faster, and hugging the old sea wall, until it burst and flung its shredded masses far into the air. The thing was altogether so realistic, and the reproduction so absolutely accurate, that it fairly astounded the beholder. It was the closest copy of nature any work of man has ever yet achieved.[21]

The great French film historian Georges Sadoul would later recall that his mother had never seen the sea, growing up in Lorraine, and was deeply impressed by the Lumières' *La mer* when she visited Paris in 1896. Several days later she went to the Breton seaside and was struck by its "resemblance" to the images she had seen at the Grand Café, so much so that "fifteen years later she still spoke of them with emotion to the small child that I was."[22] In Russia, the veteran music critic Vladimir Stasov went to see the same Lumière program with the composer Alexander Glazunov, and described its impact in a letter to his brother that ended, "And to watch the sea moving just a few feet from our chairs — Mendelssohn's *Meerstille!* — yet this silvery movement produces a movement of its own."[23] Now, with *Sea Cave*, Paul had his own maritime masterpiece, described in his catalogue as "a very striking and artistic photograph of a large cave on the Atlantic coast, into which waves dash with great violence." And this would be the first film that he took steps to copyright.

The Spanish and Portuguese films were taken during August and September, and first shown in Rousby's programs at the Real Coliseu between October and December, including two additional bullfight subjects that were withheld in Britain. Apart from *Sea Cave*, the only one of the films known today, *Andalusian Dance*, was performed by two Spanish sisters who were part of the Lisbon theater's company, and has only survived, appropriately, through its cameraman's invention, as one of Short's Filoscopes.[24] Meanwhile, the *Tour* had its British premiere on 22 October at the Alhambra. In mid-January, Rousby said farewell to his loyal audiences in Portugal, promising that he would return with more new subjects, but after a season at the Alhambra in Brighton, he appears to have returned to his earlier "electrical entertainment" act, continuing to tour widely, although not in Spain or Portugal.

Back in England, however, the *Tour in Spain and Portugal* would have a novel and unprecedented aftermath. In January 1897, the popular writer George R. Sims published a short story in the *Referee*, entitled "Our Detective Story."[25] The narrator is a private detective, recalling a visit to the Alhambra, where he sees a former client, who had sent him to Madrid to "shadow" his wife, then suspected of an adulterous liaison with his business partner. As they talk socially, the lights in the theater are lowered, and an Animatograph show begins, with the second item "A Tour in Spain."

> The pictures were wonderful, and we applauded with the packed house as the crowds moved on the canvas, each individual a living, breathing entity.
> The sixth picture was the public park in Madrid. Several people passed. Then a lady and a gentleman came on arm in arm.

I uttered a cry. So did Mr ——, the husband.
The gentleman was his partner—the lady was his wife!
Mrs —— gave a shriek. Her husband seized her by the arm. "Stay madam," he hissed in her ear. "I will see the end of this."

The husband forces both wife and partner to watch, as the couple on screen sit down on a bench and kiss, at which the audience "yelled with laughter." The husband utters "an oath of rage," his wife sobs hysterically. The story ends with a coda, "In Court," where the divorce case is being heard, and the cinematographe is shown again as evidence. The moral of the story, the narrator tells us, is to "always keep a sharp look-out for the gentleman who takes pictures for the cinematographe."

The idea of moving pictures as "evidence" of clandestine sexual activity would return in a number of early stories and even on screen, in the early American comedy *The Story the Biograph Told* (Biograph, 1904), where a husband and wife see the man's dalliance with his stenographer at a picture show, filmed by a mischievous office boy. Sims's story undoubtedly owes something to the Victorian fascination with the verisimilitude of photographic evidence, reflected in the use of "spy cameras," linked with the contemporary vogue for detective fiction. And indeed, less than a week after Sims's story appeared, an advertisement for Slater's Detective Agency in the London *Standard* offered to "carry out this new process of Photography" and to "produce the pictures in court as evidence," using the very term, Animatograph, that Paul had coined for the Alhambra shows.[26] This unusual exchange between "art" and reality offers us a rare glimpse of early moving pictures already forming part of modern city culture.

Early in 1898, Short was again dispatched abroad to film another travelogue series, this time in Egypt. His thirteen films included scenes in Port Said, in Cairo, on the banks of the Nile, and at the pyramids—but nothing, it appears, on a par with *Sea Cave*. The problem may have been that Egyptian monuments were already well known to all who attended lantern lecture travelogues or collected stereographs, which overfamiliarity seems to have steered Short toward subjects of "everyday life."[27] Cecil Hepworth, who had not yet entered the film business, openly questioned the suitability of static Egyptian antiquities for "animated photography," although two titles in this series offered unusual scenes highly suited to film.[28] The Procession of the Manunal, or Holy Carpet, was an Islamic ritual centered on the Egyptian government's creating each year a new, highly decorated cover for the Holy Kaaba in Mecca, and offering this to the Kingdom of Saudi Arabia. Traditionally it involved a large ceremony in Cairo, with a military parade visiting different parts of the city. It has been filmed on several occasions, but Short's *Procession of the Holy Carpet from Cairo to Mecca* was certainly the first.[29] The other intriguing film was simply titled *Dervishes Dancing*, and would also probably have been the first time that practice had been filmed.

Reaching the World

Queen Victoria's Diamond Jubilee, as we have seen, had been staged as a "festival of empire," and presented ideal filming conditions. Among the many companies that produced Jubilee series, Paul's would have benefited from his established reputation among exhibitors. In Melbourne, Harry Rickards promised on 13 August 1897 "an enormous attraction" in the following week, which turned out to be the Jubilee films at the Opera House that had hosted Hertz just a year earlier. The *Melbourne Herald* recorded breathlessly:

> one of the most thrilling spectacles ever witnessed, the appearance of Her Most Gracious Majesty on the Royal Carriage, drawn BY SIX CREAM PONIES, CAUSING A PERFECT BLIZZARD OF loyal and Affirmative EN-THUSIASM, the vast audience rising EN MASSE, cheering incessantly until the picture was reproduced.[30]

Details of the vast and carefully composed procession probably meant more to distant audiences than to those at home. In Melbourne, "the waving arm of Sir George Turner," the Australian prime minister, was reported to be "loudly applauded every evening."[31]

In Toronto, Paul's Animatograph had replaced the Edison Vitascope late in 1896 at Robinson's Musée, where it was "said to be much superior to the Vitascope." And even in Montreal, where newspapers appeared largely indifferent to animated photography, despite its French pedigree, the Animatographe opened on 30 September 1896, with shows "every afternoon and evening" on St. Catherine's Street; then in November audiences for the Theatroscope could see "the Prince of Wales' horse win the Derby" in Paul's now famous film. When the Diamond Jubilee films reached Canada, the dominion's first premier of French Canadian ancestry, Wilfrid Laurier, seen surrounded by contingents of Canadian troops, was appreciatively recognized by local audiences in both communities, who knew that he had been knighted on the morning of the Jubilee pro-

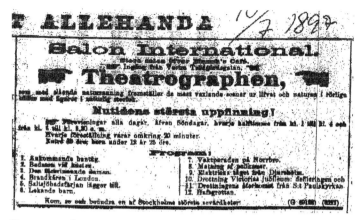

40. Newspaper advertisement for a screening of Paul's films in Stockholm, Summer 1897.

cession. But the full extent of film's potential for creating a new imperial connectivity and empathy would not be revealed until two years later, during the Anglo-Boer War.[32]

Having taken an active part in filming the Jubilee procession, Paul seems to have treated himself to a foreign filming expedition. Nearly forty years later, he explained the circumstances of his visiting Sweden in the summer of 1897:

> The King of Sweden and Norway sent his artist for a projector, with instructions that the maker was to accompany it and see it properly installed in the Palace at Stockholm. This I did, I hope to his satisfaction, and I was granted special facilities for getting Swedish pictures.[33]

What he does not explain in this laconic account is that there was every reason for him to want to be in Sweden during the summer of 1897, as its capital was hosting the Stockholm Art and Industry Exhibition (*Stockholmsutställningen*), the latest in the series of world's fairs that had started with London's Great Exhibition in 1851. Even more to the point, one of the special themes in Stockholm was new media. King Oscar's opening speech was recorded on phonograph, and his arrival formed part of a Lumière presentation, the Kinematografen, a twenty-minute program of subjects that spanned the world. Röntgen's X-rays were also on display, along with new developments in telephony.[34]

If Paul regretted that his equipment was not featured in the exhibition, this may have spurred him to produce a series of subjects that reflected its dual focus: Swedish tradition and celebrating new technology. Part of the display was a re-creation of traditional life in the Skansen open-air museum, where Paul filmed three version of Swedish national dancing, and probably two scenes with Laplanders. His *Electric Trolley Car* was probably the Skansen Bergbana, a new funicular railway, and another film showed derrick cranes coaling ships. Paul's films were shown at a Salon International, as part of a café in the King's Garden in central Stockholm, and the run appears to have been extended across the summer due to demand. Paul must have trained one or more local operators to maintain this show, as he had already done in London. Its contents would presumably have been similar to Rousby's Spanish and Portuguese programs, with a mixture of local films, which were proving attractive throughout the world, and films of famous foreign places—in this case Britain, and specifically London. What was local in one market could obviously become exotic in another, rendering films of identifiable "place" attractively versatile. To make them so, however, required showmanship, about which there is little evidence, apart from the anecdotal and what can be gleaned from playbills and posters.

Travel by Film

Some idea of changing expectations for actuality programs can be deduced from the language and style of posters between 1896 and 1900. A Lumière playbill of 1896 for the Empire Theatre Cardiff listed films of Moscow, Hamburg, Madrid, Paris, Stuttgart, Barcelona, and London, as well as "Burmese Dance at the Crystal Palace" and "Tobogganing in Switzerland." Potential customers were solemnly reminded that they would have to "expend a large amount of money and time to obtain a view of the Scenes of the above Programme in their Geographical situation," whereas they could do so economically merely by visiting the theater. Paul's own program at the Alhambra in London on 31 August 1896 comprised at least twenty subjects, of which six could be described as dramatic or performances, four topical, and the remaining ten actualities. By 1898 a playbill for a fundraising event in Cheltenham that Paul was supporting made an effort to link the disparate films in a kind of touristic narrative:

You can MEET THE PARIS EXPRESS, at Calais, to see THE FIRE BRIGADE, called out. Then back to WESTMINSTER, take the first 'bus for PRINCESS MAUD'S WEDDING; after that to HAMPSTEAD HEATH in

time to see TOMMY ADKINS' COURTSHIP, then with the crowd to THE
DERBY. . . . The GORDON HIGHLANDERS pass you on the way to the
MUSIC HALL SPORTS, and many other events.[35]

Two years later, the ebullient "Edison" Thomas would intersperse his extended programs with such items as "HOLIDAY SCENES at Blackpool,
Morecombe and other Popular Resorts," the ever popular "Turn-out of
the Fire Brigade," "Scenes at the Cod Fishery," and two "trip" sequences,
"A Trip to Newfoundland" and "A Trip to the Alps."[36]

Paul's hectic rate of new releases continued, with eighty titles released
in 1897 and slightly more in the following year. A continuing emphasis
was on series of actualities devoted to British locations, including the
seaside resorts of the Isle of Man and Brighton, as well as street scenes
in various cities: London, Leicester, Sheffield, Burnley, Hull, Liverpool,
Cardiff, Glasgow, and Edinburgh. The only foreign series from this period
was three subjects from Constantinople, before the South African war
began to dominate Paul's output at the end of 1899.[37]

Animatography and Ethics

Just over a year after the Bazar de la Charité fire in Paris, London experienced its own disaster on the River Thames, with film once again impli-
cated. On 21 June 1898, the battleship HMS *Albion* was being launched
by the Thames Ironworks and Shipbuilding Yard at the junction of Bow
Creek and the Thames, usually known as Blackwall. This largest yard on
the Thames had developed a worldwide reputation for warships after
building the world's first iron-hulled frigate in 1860, attracting orders
from a number of countries, including two battleships for Japan immediately before the *Albion*. Large crowds had gathered to see the Duchess
of York, later Queen Mary, perform the ceremony, and no fewer than
four film companies were present to record it. The duchess succeeded in
breaking the champagne bottle only on the fourth attempt, but when the
ship slid toward the water its backwash swept away an old wooden bridge
crowded with spectators, plunging hundreds into the river.

Birt Acres was the official filmmaker, licensed by the shipbuilder and
mounted high on the scaffolding, E. P. Prestwich filmed in both 35- and
60-millimeter formats, and the German company Philipp Wolff would
also offer a film of the event.[38] For Paul, the launch of the *Albion* seems to
have been intended as part of an extended series of maritime subjects. In
February, he issued *Launching a Liner*, which showed the launch of the
Braemar Castle on the Clyde, with "the vessel appearing to glide down

41. *The Launch of HMS Albion* (1898). On 21 June, over thirty thousand people gathered to watch the launch of the battleship from the Thames Ironworks at Bow Creek. The wave created when the ship entered the water swept away a flimsy wooden stage, sending hundreds of spectators into the water, with an eventual death toll of thirty-four. Paul helped rescue victims and donated proceeds from showing the film to charitable help for the bereaved.

towards the audience." (Two years later, this and two other ships filmed by Paul embarking passengers, the *Dunvegan Castle* and the *Hawarden*, would be carrying troops to South Africa.) The theme continued in March with *Unloading Cargo at London Docks* and two "panoramic views, taken from a tug," of shipping on the Thames, with a view of the Embankment and Cleopatra's Needle no doubt taken on the same trip. So Paul already had experience filming on the river when he hired a tug again in June, and took the opportunity to make a double-length film of "an enormous steam floating derrick, coaling several steamers at once," the latest stage of William Cory's massive coal-loading operation at Victoria Dock, which had driven London's industrial growth with cheap coal imports since the mid-nineteenth century.[39]

Having filmed the *Albion* reaching the water, Paul must quickly have found himself surrounded by onlookers who had been swept into the river. The surviving second part of his film, code-named "Disaster," shows many small boats milling around the riverside scaffolding, people in the water, and a man gesturing vigorously toward the camera, directing Paul's boat either to leave or to go elsewhere. A woman caught in close-up on the boat could be Ellen Paul, who may have come to enjoy the spectacle of the royal launch. But the fact that the disaster had been filmed would produce a storm of controversy. Birt Acres was the first to write, somewhat priggishly, raising an ethical objection to the event being displayed:

It having come to my knowledge that someone had taken an animated photograph of the poor sufferers struggling in the water, I wish to dissociate myself in the most emphatic manner possible from the producers of these photographs; and further, I have decided to suppress my films of the launch.[40]

No friend to Paul since their acrimonious split, Acres was clearly referring to his "Disaster" film. Another anonymous correspondent wrote the following day "with deepest sorrow" to deplore that "the scene of the tragedy of Blackwall was exhibited by means of the 'living picture' apparatus before an audience at the Royal Music Hall, Holborn, on Wednesday evening," adding that "surely so grave an insult on the best feelings of mankind ought not to pass unreprimanded."[41]

In fact, the Holborn screening provoked a display of respect as unprecedented as the "living picture" itself, as the *British Journal of Photography* reported:

The whole audience rose with bared heads as the drama of death passed before them. At the close, with the orchestra playing "Rocked in the cradle of the deep," there was scarcely a dry eye among the silent audience.[42]

Another journal, the *Amateur Photographer*, described by John Barnes as "ever a devoted supporter of Acres," echoed the "feeling of revulsion" that made Acres stop filming, and reported his speaking at a photographic club about the "less delicately-minded individual" who had filmed "the actual drowning event."[43] Paul would emphasize in a letter and in his advertisements that his launch had "picked up twenty-five of the submerged persons," having been "beckoned towards the fatal spot," and that "no consideration of [filming] was allowed to interfere with the rescue."[44] He went on to record that he had helped organize other music-hall screenings, with "a collection to be made for the sufferers"; also that he was "sending it to local exhibitors throughout the country," hoping that "the sympathy and interest thereby excited will result in additional subscriptions for the distressed families."[45] Another photographic journal, the *Photogram*, avoided taking sides in this war of sensibilities, while reporting the salient fact that Paul's camera was electrically driven and could therefore run without an operator's attention while rescue work was under way. The *Photogram* also noted that Acres was too far away from the river scene to have been able to help—or indeed for his film to have shown any detail.[46]

Knowing as we do the extent of Acres's animosity toward Paul, it is tempting to read this controversy mainly in ad hominem terms. From a modern perspective, it is easy to see Paul as being both pragmatic, in recognizing that he had a scoop, and charitably practical, in seeing the potential for raising money for victims' relief. By contrast, Acres had little to lose in protesting his sensitivity and "suppressing" film that would in fact have shown little of the drowning tragedy, while also relishing an opportunity to embarrass Paul. But beyond this sparring, significant issues were raised, probably for the first time, about the status and propriety of the filmic record. Photographic media produced what would later come to be known as "indexical" images, equivalent to an imprint of their subjects, which caused many to feel uneasy about their transgressing taboos surrounding death.[47] The first press reviews of the Lumière Cinématographe in Paris in 1895 already referred to the new phenomenon's challenge to mortality.[48] And early in the new century, writers from very different traditions would reflect on the shock of seeing long-lost loved ones, or indeed the act of killing in lifelike form.[49] But perhaps the most contentious aspect of film of the *Albion* disaster was its being shown in vulgar places of entertainment—almost as if the corpses of the *Albion* victims were on display. After the outbreak of the South African war, a year later, Paul would undoubtedly have given considerable thought to how war and its victims might be acceptably shown on screen.[50]

Screen Tourism

By the time of Queen Victoria's funeral in February 1901, many taboos in representation had been breached, and there was no question that the funeral cortege would be recorded by Paul and the other British film companies, to be distributed throughout the empire. Soon the succession of her son Bertie as Edward VII would provide an opportunity for the most exotic territory of the empire to become visible on British screens. Edward had no more intention of traveling to India to be proclaimed emperor than had his mother when she became empress. But just as Joseph Chamberlain's Diamond Jubilee procession had created a magnificent "imperial" spectacle in London, so the new king's symbolic installation in India could become an impressive filmic event.

Lord Curzon had been appointed Viceroy of India in 1899, and although the crown was only to be represented by Prince Arthur, Duke of Connaught, to mark Edward becoming Emperor of India he organized a vast durbar, or gathering of Indian nobility, early in 1903. Among its most spectacular features was a parade of 150 richly decorated elephants and a march-past of thirty-seven thousand troops under the command of Lord Kitchener, both of which provided excellent subjects for film-ing. Paul issued at least five films based on the durbar, including one of a polo match between Chitral and Gulghit provinces, and another of the Duke and Duchess of Connaught visiting the Khyber Pass. Within a few months, more films of royalty and of big-game hunting in India followed, presumably to capitalize on interest aroused by the durbar.

Amid this imperial splendor, which was even more lavishly recorded by Charles Urban's rival Warwick Trading Company, Paul returned to his interest in Scandinavia, with an extended series of eighteen films taken in Norway covering its scenic delights from Bergen to Lapland, and in-cluding a railway journey from Vossevangen to Bergen. Railway scenes had remained popular since film's earliest days, and during 1903–1904 Paul would issue a variety of British rail journeys, including *Derby to Bux-ton, Scenes on the Great Northern*, and *Panoramic View from the Scottish Express*. The climax of these was with a fourteen-part series, *By Rail to Penzance*, which included as scenic highlights Roman ruins at Bath, the Clifton suspension bridge, the "rugged Cornish coast" at Land's End, and the Scilly islands. Given the reference to specific companies in the titles of some of these, it is possible this was a commission from the long-established Great Western Railway Company, which was active in pro-moting tourism in the southwest. Paul may also have been encouraged to produce this extended series by the launch that year of Hale's Tours of the World, a novelty entertainment format that seated its audiences in a railway-style auditorium, with film shot from the front of a locomotive giving the illusion of travel through exotic scenery.[51] Strange to relate, this proved genuinely popular, and remained a successful attraction on London's Oxford Street until 1910.

Even for audiences in more conventional situations, the connection between film and travel was firmly established. Watching a film did "trans-port" viewers to remote places and communicate the sense of journeying. It also gave audiences a sense of "what to see" and, in an era of expanding amateur photography, what to photograph. The number of nonspecialist hobby photographers had been growing rapidly during the 1890s, encour-aged by a series of cheap cameras marketed by the Eastman Kodak com-pany, culminating in the Brownie in 1900. Paul must have been aware of this, and presumably realized that his experience filming around London over five years could be used to create some beneficial promotion of his company's name. Hence his 1901 "folding card" brochure entitled "What to photograph in London."[52] This must have been well-received, since it was promoted in a more mainstream periodical in 1903:

Mr. R. W. Paul, the well-known manufacturer and dealer in scientific and photographic apparatus, London, has arranged the following useful list of places photographed in London. For the convenience of visitors it is arranged in six groups, each group representing a day's work or more, according to the zeal of the amateur. Interiors have been omitted, architectural photography requiring more time than is usually at the disposal of the visitor; but further information for those wishing to take churches, palaces, etc., will be afforded on inquiry of Mr. R. W. Paul.[53]

Paul was of course a Londoner who had lived and worked near its center all his life, and some of his earliest films in 1896 showed the city's most famous thoroughfares, especially its bridges. But Roland-François Lack observes that in its detailed itineraries "there is little overlap between the selection of sights he makes for photography enthusiasts and the London that appears in his films."[54] In fact, Paul's commentary narrativizes the process of gathering popular images, anticipating the "guided tour" format of later film series, such as *Wonderful London*.[55]

New Trends in Nonfiction

Throughout the first seven or eight years of the new century, there was still a market for nonfiction films, although it was becoming highly competitive. Charles Urban was specializing in nonfiction with his Urbanora brand, promising to "put the world before you."[56] In 1904 Paul began to use a "day in the life" format for some of his actualities, with subjects ranging from coal-mining to racehorse training and railway operation.[57] Until recently, none of these was available to view, but the 2016 discovery in the Swedish Film Archive of *A Collier's Life* reveals this to be a pioneering account of "how coal is won and despatched from the colliery." Filmed at Shirebrook Colliery in Derbyshire, then one of the most technically advanced in Britain, its five scenes originally covered checking the Davy safety lamp, "holing" at the coal face, moving loaded tubs, and sifting the resulting excavation heaps, with a lunch time scene of miners amid the pit's considerable infrastructure. Another film from the same year, *How Our Coal Is Secured*, may have been taken on the same occasion, recalling that as an engineer, Paul remained interested in technology and industrial processes. *A Collier's Life* may have been filmed by his then employee J. H. Martin, who would go on to make what has long been considered one of pioneering examples of British documentary, *A Visit to Peek Frean & Co's Biscuit Factory*, just two years later.[58]

42. *A Collier's Life* (1904). This prototypical "industrial" film—discovered and restored by Camille Blot-Wellens at the Swedish Film Institute archive in 2016—shows
Shirebrook Colliery in Derbyshire, which had opened in 1896 and gained early notori-
ety for strikes over poor living and working conditions. Paul's film may have been part
of an attempt to improve its image. Courtesy of the Swedish Film Institute Archive.

Topical films were apparently still important and profitable, to judge
by a detailed account in the trade press of Paul's manager, Jack Smith,
and his team at work in 1905.[59] Headed "Something like enterprise," the
report described how a film of the royal review of Scottish Volunteers in
Edinburgh was taken and distributed to Scottish exhibitors with the same
urgency Paul had managed with his original Derby film of 1896:

Accompanied by members of his darkroom staff [Smith] travelled all
Saturday night arriving at 7.30 on Saturday morning, and after breakfast
immediately proceeded to fit up the dark-room. Water had to be laid on,
developing tanks erected, printing and perforating machines fixed, and
the most important item of all—the drying room was arranged, where
special heating apparatus had to be fixed. Everything was ready by seven
o'clock Sunday night. The call being for six on Monday morning, certain
members were despatched to take up their positions at the Royal Review,
and others were sent to the railway station to take the King's arrival, all
returning as soon as possible to the darkroom after the Review, and work
was immediately commenced. By persistence and no stopping for meals,
five prints were shown the same night; two in Glasgow, one in Dunferm-
line and two in Edinburgh. This, however, was not the finish of the work

owing to numerous orders coming in, and all had to work until 9 o'clock on the Tuesday morning, when after 27 hours' hard work all retired to their bed thoroughly tired out, and after three hours' sleep they went at it again.

This clearly promotional account implies that there was good business to be made from delivering a film of King Edward traveling north of the border. And it ends by observing that "Mr Paul and his manager have their eyes open for topical events, and do all they can to look after their clients, and help them keep their exhibits thoroughly up to date."

The only surviving actuality from this period is also Scottish and royal, although something of a conundrum. *Aberdeen University Quatercentenary Celebrations* records a royal visit to Aberdeen in October 1906, when Edward VII opened a new building of the historic Marischal College. But although this is the longest surviving film attributed to Paul, at over thirty minutes, there remains some doubt about its origin. According to a contemporary trade journal, it represented "smart work" by Jack Smith and three assistants, who traveled to Aberdeen by train, shot the film, and processed it back in London, before sending it for screening "by a local exhibitor" on 29 October before a distinguished audience.[60] However, according to local publicity, it was made by Walker & Co of Aberdeen, and presented as "the longest cinematogram ever taken in the world," at fifty minutes.[61] The Scottish account identifies two local cameramen, making no mention of Smith or Paul. What seems likely from the leisurely surviving version is that there may indeed have been up to fifty minutes of material, which could have benefited from editing for nonlocal audiences; and possibly Paul had been contracted to help Walker bring off the largest production of their career.

The last surviving nonfiction film produced by Paul, from 1908, offers a view of an industry that few today realize once flourished around Britain's northern coasts. *Whaling Afloat and Ashore* belongs to a group of Paul releases filmed in Ireland, which may have been taken by the same cameraman.[62] Its opening section, filmed on a whaler in rough seas, still seems distinctly innovative. Through six minutes, from the title "First sight of a whale," we see "a successful shot" and "the tethered whale blowing," then a second harpoon fired at close range, enabling the whale to be towed to land. The film's remaining six minutes are devoted to the various processes of cutting up and extracting materials from the carcass, including whalebone and oil. Finally, the men of the whaling station, identified as "Norwegian and Irish" are seen "at play": dancing together, wrestling, and tossing a boy along a line.

43. *Whaling Afloat and Ashore* (1908). The latest surviving example of Paul's longer films documenting industrial processes—in this case the whaling industry established by Norwegian companies on the West coast of Ireland between 1908 and 1914.

The film may well have been commissioned, or at least prompted, by a Norwegian company's establishing the Arranmore Whaling Company in 1908, based on South Inishkea, a small island off County Mayo in the northwest of Ireland. Arranmore was apparently successful, catching nearly four hundred whales between 1908 and 1913, which led to another Norwegian-owned company setting up nearby. Despite the First World War interrupting their activities, these enterprises continued until 1922, and the hunting of whales in Irish waters lasted until an act of 1976 made it illegal.

The final year of Paul's film production, 1909, saw only one major non-fiction release, *The Newspaper World from Within*, which at 1,200 feet would have been his longest film (apart from *Army Life* and the *Aberdeen Quatercentenary* series). Apparently continuing the documentary impulse of earlier films, it dealt with the *Morning Leader*, which had been one of the first English newspapers to publish a daily edition; presumably the film covered how this was achieved. The *Leader* may at the time already have been in some difficulty, as it was bought by the Quaker industrialist and temperance campaigner George Cadbury and merged with the *Daily News* in 1912.

Looking back at Paul's extensive nonfiction production, we might wonder if the declining demand for such subjects was one of the factors contributing to his decision to leave film in 1910. The composition of film programs was changing by then, as exhibitors began to show a preference for longer fiction film, soon to be distributed under "exclusive" contracts. But in 1907, opinion about future trends in exhibitor demand was clearly divided. In an early issue of the trade journal *Kinematograph and Lantern Weekly*, the London-based American film producer and entrepreneur Charles Urban declared that "the entertainment side of the business has now reached its maximum . . . and future development will be upon the educational side."[63] Urban was already distributing large quantities of nonfiction, and would continue to do so, adding distinctiveness to his output with the promotion of Kinemacolor the following year. But just a month later, the editor of this new journal, Theodore Brown, offered two ideal or "typical" programs for the guidance of exhibitors, which he assumed should contain "suitable proportions of contrasting items," listing these as "comics, panoramas, trick and dramatic subjects."[64] Of the seventeen titles in one of Brown's programs, five can confidently be identified as nonfiction, while a mere four appear in the second; only one (fictional) title from Paul appears in the two lists, the 1906 comedy *Undressing Extraordinary*. In the same issue of *Kine Weekly* (as the journal soon became known), a listing of "Latest Productions" from eleven producers includes 150 titles, of which approximately 40 can be identified as nonfiction, while Paul's contribution of 12 titles includes only one nonfiction subject (*Sheep Shearing*).

From this snapshot of the industry about two years before Paul left it, it seems clear that demand for the short "interest" or "topical" film was declining, with only the most topical or spectacular holding their place in average programming. And by 1909–1910, the number of titles required for a normal program had been virtually halved, as the most attractive films became longer. Analysis of a typical program from a successful provincial cinema in mid-1909 shows that the program required only nine films, of which two were actualities.[65] Yet there is also evidence from this period of Paul remaining closely involved with his production business, and especially with "topicals." A report from September 1908 described him with Jack Smith "at Euston [the London station] saying 'goodbye' to a representative from Dublin on the 10.15."[66] While there, Paul and Smith saw a headline in the night edition of the *Star* about the wreck of a ship off the South Wales coast:

Mr Paul: "I think it is worth while running down"; and Mr Smith at once replied "Yes, I will go back in the car to Southgate," and away he rushed

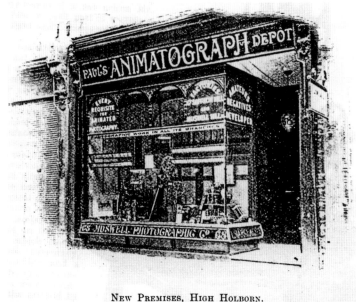

NEW PREMISES, HIGH HOLBORN.

44. Paul's Animatograph Depot showroom on High Holborn. This appears to have remained the retail base of his film business, separate from the concentration of other early film companies in Cecil Court, off Charing Cross Road. Photograph from Paul's 1900–1901 catalogue, *The Hundred Best Animated Photograph Films*. Courtesy of the National Fairground and Circus Archive, University of Sheffield.

to get his camera, tripod and paraphernalia, and got back to Paddington in time to catch the one a.m. train, which arrived [in South Wales] at 6.30 in the morning.

Like the previous report of Smith's devotion to deadlines, this goes on to record in detail his perseverance in filming "what was left of the barque *Amazon*," as well as pictures of the three sailors rescued, before catching the four o'clock train back, "in order that Londoners might witness the pictures almost as soon as they read the news."

What was beginning to emerge, near the end of Paul's film career, was a prescient realization that print and film media could work together to deliver news—at the same time that the idea of the film "newsreel" was emerging.[67] Equally interesting is the revelation that Paul was personally involved in his film business as late as September 1908, to the extent of accompanying Smith to Euston for a social farewell. And *Kine Weekly*, which had hailed *Whaling Afloat and Ashore* as "a film which deserves to rank as a headliner on any bill," was enthusiastic about the promotional strategy of distributing "a dainty little postcard, one side of which bears four views from his recent great success."[68] In all of these strategies, we find Paul continuing to innovate and optimize, just as he did in the instrument business, in this even newer industry he had played a central role in creating.

CHAPTER 7

Distant Wars

SOUTH AFRICA AND BEYOND

On 11 October 1899, the autumn of the last year of the century was rudely interrupted by two South African Boer republics, Transvaal and the Orange Free State, declaring war on Britain. Their demand, that Britain withdraw its forces from their borders after a series of provocations and threats, was inevitably ignored. Indeed, response to the ultimatum in Britain was largely contemptuous, with the *Times* calling it an "extravagant farce" and the *Daily Telegraph* declaring, "Kruger has asked for war and war he must have."[1] Like many disastrous conflicts, this second Anglo-Boer War began with rejoicing.[2] The stakes were considered high: in a speech on the evening of 11 October, Herbert Asquith bluntly asked, "Is Great Britain the paramount power of South Africa?," and a recalled Parliament voted overwhelmingly to fund a major military response.[3] Three days later, General Sir Redvers Buller left Southampton with the advance party of an expeditionary Army Corps, intending to teach the upstart Boers a lesson.

Would films play any part in connecting Britain and its far-flung empire with this distant war? There were already signs that they might. The 1898 Spanish-American War, initially over Cuba, had produced a surprising number of films, while in that same year, Paul made two films about the rapturous reception given to General Herbert Kitchener after his victory in the Battle of Omdurman in Sudan.[4] In responding to public interest in a war, early filmmakers were following the tradition long practiced by printmakers, lantern-slide and sheet-music publishers, and of course popular stage performers.[5] With a new war begun, Paul and other British filmmakers responded almost immediately. Yet until recently, film received little mention in accounts of the second Anglo-Boer War. Thomas Pakenham's landmark history, *The Boer War* (1979), contained a solitary mention of the intrepid W. K. L. Dickson "dragging his enormous ciné camera" up a hill overlooking the Tugela River.[6] But Pakenham was writing before the emergence of a new archaeology of early film, and had almost certainly never seen any of the more than sixty films Dickson shot in Southern Africa during 1899–1900 for the Biograph and Mutoscope Company, or known how prominently they featured in programs at the company's flagship Palace Theatre in central London. Twelve years later, John Barnes boldly titled the fourth volume of his history of early British film, covering 1899, *Filming the Boer War*, claiming that the war "had a profound effect on the cinema in Great Britain," with "the demand for new subjects depicting every aspect of the conflict" and "an interest in pictures of the war shared by all classes" resulting in "a boom for the industry."[7] In terms of quantity, Barnes was undoubtedly right: at least five hundred war-related films were produced over three years.

Building on this new awareness, Simon Popple has collected contemporary evidence of "a positive craving . . . for pictures dealing with the war."[8] Not all of these would be moving pictures—stereographs clearly offered a closer and more vivid sense of the terrain and those fighting in it.[9] But even if there is now a risk of overestimating film's status, or even its visibility, the quantity and variety of production is undeniable. And no producer was more inventive than Paul, who devised at least five different kinds of war film during the first twelve months of the conflict. He also acted fast, sending two film cameras to the Cape just a week after Biograph's Dickson had embarked with Buller and the British advance party.

The confrontation between the South African Boer republics and Britain had been brewing for a decade and was unusual among imperial conflicts. Ostensibly Britain was supporting the rights of the *uitlanders*, or non-Boer settlers in Transvaal and the Orange Free State. After the discovery of gold in Witwatersrand in 1886, predominantly British immigrants had poured into Transvaal, and in 1890 legislation was passed to deny them voting rights. The Jameson Raid of 1895, organized by Cecil Rhodes and covertly supported by Britain, was intended to spark a rising among the *uitlanders*, which failed to happen. When the Boer leaders seemed to have conceded to Britain's demands in the summer of 1899, the newly arrived British high commissioner, Alfred Milner, was frustrated. Yet his superior, the foreign secretary Joseph Chamberlain, encouraged confrontation, writing on 7 October that he believed "Milner and the military authorities greatly exaggerate the risks of this campaign."[10] On the other side, one of the rising figures in Afrikaner politics, Jan Smuts, foresaw that "South Africa stands on the eve of a frightful blood-bath," which might lead either to servitude or to "an Afrikaner republic stretching from Table Bay to the Zambezi."[11] The scene was set for a protracted conflict between resourceful opponents, separated from Britain by a three-week voyage.

If Biograph made the biggest and earliest commitment to the war, both Warwick, led by Charles Urban, and Paul acted almost as quickly. Paul sent the first of two cameras with Colonel Walter Beevor, a medical officer attached to the Scots Guards, when they embarked in the SS *Nubia* on 21 October. Compared with life on Buller's ship, the *Nubia* was described by one of the guards as

> not a great troop ship as such is now, the food was poor, often the ration included ships biscuits, probably the hardest of its kind ever baked, there was also pickled pork at times, a very unpalatable dish. . . . The only incidents of the voyage was a stop at St. Vincent and the ceremony of crossing the Equator, with a fairly bad bit of weather near the Cape. We disembarked on the 14th November, recovered our good name and entrained at once for Orange River where we prepared for the advance.[12]

The Boer commanders also acted fast after the rejection of their ultimatum, raising militia commandos that were highly mobile and armed with newer equipment than the British army had. After capturing a small British garrison on 12 October, within days they besieged the major towns of Ladysmith, Mafeking, and Kimberley, taking advantage of the delay before British reinforcements arrived.

British confidence quickly faded in the face of this devastating news. On 14 October, an Islington manufacturer had advertised waxwork figures of Transvaal president Paul Kruger "modelled from the very latest photographs," presumably to feed popular scorn for the Boer leader.[13] But on the last day of the month, coincidentally when Buller reached Cape Town, the most popular poet of the day, Rudyard Kipling, published his ballad "The Absent-Minded Beggar," reminding his readers of the hardships that would face "the gentleman in khaki ordered South."[14]

It is difficult to judge what part films played in public perception of the war, except to note that this was the first conflict involving Britain and its extended empire to be extensively referenced on film. The earliest film programs including war-related subjects, at Biograph's Palace Theatre, benefited from Dickson's early departure.[15] Films of Buller's departure appeared in Palace programs as early as 19 October, and by 21 November his arrival at Cape Town was also screening. "Departures and arrivals" were indeed the first major war genre, and Buller's progress from London via Southampton to the Cape was not the only journey covered in detail.[16] In Paul's 1900–1901 catalogue the Transvaal War section is sub-

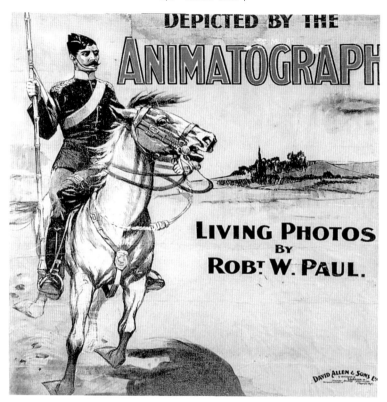

45. This large colored advertisement for films from the Anglo-Boer War (1899–1902) is the only known example of a poster produced by Paul, no doubt occasioned by the demand for films about the war. National Science and Media Museum, Bradford.

divided into four groups, beginning with "Leaving for the Front." The first film listed emphasized the individuality of the troops:

> This fine picture shows the Scots' Guards (one of the first to go) leaving for the front, as they are embarking on board SS *Nubia* at Southampton. The men face the camera as they pass up the gangway, and a clear picture of each man's features is obtained.[17]

The film is lost, so it is impossible to know if its unusual view of men facing the camera was due to its being taken from the ship, perhaps by Colonel Beevor.

The "tour" format of presentation, often still reliant on lantern slides, with a minimum of films, was also quick to include references to the war. In the North London suburb where Paul was now based, one such show took place on 4 December. Billed as "A Trip to the Transvaal" and "illustrated by cinematograph and oxy-hydrogen lantern," this church hall event was actually an account of a prewar South African tour that had now acquired topical interest, with the result that "every seat was occupied long before eight o'clock and late-comers were lucky if they obtained standing room."[18] The lecture began with a group of films showing the launch of the *Braemar Castle*, the embarkation of passengers, departure, and games on board, followed by slides covering the voyage and arrival at Cape Town. Two films of Cape Town showed the modernity of its electric trams, balanced by the more traditional image of a mounted policeman "cantering down the street ... received with applause." Finally came a sequence of slides ranging over Kimberley, Johannesburg, the Modder River, and various places "where a battle has since been fought." All of the films shown appear to be from the Warwick Trading Company's extensive catalogue of actualities.[19] The evening also included songs performed by the lecturer, typical of the church-centered social culture of the time.[20] Similar programs took place throughout the country. In Yorkshire, Luscombe Searle presented a "celebrated 'chat' ... illustrated with songs, the latest cinematograph and limelight views from the seat of the war,"[21] while the Tees brothers advertised "Animated Photographs taken at the Seat of War," a similarly mixed film and slide show, at the YMCA in Brighton in 1900.[22]

The war was already reflected in the programs of music halls, which had long fostered popular patriotic sentiment. The Holloway Empire music hall had opened on Holloway Road in North London just three months earlier, and a regular item on its program was "Walter Gibbons's American Bio-Tableaux." Gibbons was an important early supplier of moving-picture shows to the halls, and an occasional filmmaker, who later ran the London Palladium and was knighted. Details of the Holloway Bio-Tableaux program were not normally published, but for the week of 18 December "animated photographs of war scenes" had pride of place, just below Miss Lottie Collins "of 'Ta-ra-ra-boom-de-ay' fame."[23] The twenty films listed formed a travelogue-style narrative, a "journey to South Africa"—although "war scenes, episodes and incidents" were promised, only three films actually fitted this description.[24]

A series of British defeats and continuing sieges were taking a heavy toll by year's end, reaching a climax in the Battle of Colenso on 15 December, when a Boer force of only eight thousand prevented Buller's army of twenty-one thousand from crossing the Tugela River to relieve Ladysmith and inflicted significant casualties. We get some sense of the prevailing mood from an end-of-year editorial in a North London local paper:

It is not with the usual feeling of quiet satisfaction that we approach the Christmas of 1899. There was no war cloud on the horizon twelve months ago, but we have now plunged into the hurly-burly and it is impossible to say when we shall emerge from it.[25]

One consolation was the response from elsewhere in the empire. New Zealand was the first colony to send troops, eventually committing over six thousand volunteers, who were soon joined by some twenty thousand Australians and more than seven thousand Canadians. Although supporting the mother country was widely popular in both dominions—whose settlers could no doubt identify with the English-speaking *uitlanders* in the Boer republics—there was also opposition from ethnic groups with a tradition of anti-imperialism, such as French-speaking Canadians in Quebec and Irish-Australians.

Even with the empire's support, Britain needed yet more troops than its underfunded regular army could provide. Volunteer companies began to form, among the first being the City Imperial Volunteers, a thousand-strong force recruited and funded by the City of London, which, according to Eric Hobsbawm, "reflected the appeal of patriotic propaganda" for the middle class.[26] Among the CIV volunteers were Robert Paul's own younger brothers, Arthur and George, both of whom were then working in their father's shipping business.[27]

Paul later recalled that he received usable film only from the camera entrusted to Colonel Beevor, who was "able to get about a dozen good films," including one of the war's few actual scoops: a shot of the Boer general Piet Cronjé after he surrendered to Lord Roberts at the Battle of Paardeberg on 27 February 1900.[28] This surrender marked an important turning point in the war from the British perspective, bringing to an end the series of humiliating defeats. Ending the siege of Kimberley was quickly followed by the relief of Ladysmith on 28 February, and eventually by breaking the siege of Mafeking on 17 May. Paul's contribution to celebrating this reversal began with the Cronjé film (which survives in truncated form), proudly described in his catalogue:

> This historical film, which is the only one of its subject taken, shows Cronje in a cart after his defeat at Paardeberg, followed by an escort of CIV. As the cart passes the camera, Cronje is seen to look out in astonishment at it.[29]

Building on the sense of occasion, the catalogue added that "the picture is most successful, considering the circumstances under which it

46. *Cronjé's Surrender* (1900). Boer general Piet Cronjé surrendered in February 1900, and this film of him being transported in a Cape cart illustrates the difficulty of getting close enough to take dramatic images, faced by Paul's and other cameramen covering the war.

was taken in the early morning." Paul's second "victory" film, released as *Scots' Guards Triumphal Entry into Bloemfontein* on 14 April, showed the kilted guards marching into the capital of the Orange Free State a month earlier, led by their pipers, with "almost every detail of the men's battle-stained uniforms" visible as they enter the Market Place. The marching ranks do indeed still convey what must have been the film's message to viewers across the empire — that imperial forces were reasserting control. Unsurprisingly, there is no corresponding record of the large concentration camp subsequently established by British commanders near Bloemfontein, which would become notorious for its ill-treatment of women and children.[30]

When the spring successes reached their climax with news of the relief of Mafeking on 18 May, provoking ecstatic nationwide celebration across Britain, Paul produced what was probably a staged film, code-named with the new word that had entered the language, "Mafficking":

A ragged regiment of youngsters bearing flags and beating drums are seen marching round the lions in Trafalgar Square on "Mafeking Day." Their smiling and happy faces show distinctly, and some youthful nurses bring up the rear.[31]

This was not the first time Paul resorted to staging a film to meet the topical expectations of audiences. Before the first material arrived from the Cape, he later recalled,

47. *Scots Guards Triumphal Entry into Bloemfontein* (1900). One of the films taken for Paul by Colonel Walter Beevor, a medical officer with the Scots Guards, who captured Bloemfontein in March 1900.

representations of such scenes as the bombardment of Mafeking and the work of nurses on the battlefield were enacted on neighbouring golf links, under the supervision of Sir Robert Ashe, an ex-officer of Rhodes' force.[32]

Paul's studio and laboratory in Muswell Hill were bordered by a new golf course that had opened in the developing suburb in 1894. The open expanse so conveniently close provided an ideal space for simulating the African veldt, and Paul staged at least eight "Reproductions of Incidents of the Boer War" during November and December 1899.[33] *Shooting a Boer Spy* showed the spy refusing a blindfold while facing a firing squad, while *Attack on a Picquet* claimed to be a "remarkably natural reproduction of an actual occurrence," in which a British outpost is overrun by Boers who club and shoot the soldiers, before they "rob their bodies and carry off their arms." In a more heroic vein, *The Bombardment of Mafeking* had soldiers continuing to play cards and "jeer as shells burst around them," and was described as "very funny." In *Capture of a Maxim*, British soldiers gain a heavy machine gun from the Boers. And *The Battle of Glencoe* showed the "British storming the hill and driving the Boers over the ridge with a Maxim and a strong rifle fire." *Wrecking an Armoured Train* claimed to offer "a graphic and complete reproduction of the armoured train incident at Mafeking," in which "the British are seen defending the train" and being wounded until "at last the British officer hoists a white flag in token of surrender." Of the two films known to have survived, *A Camp Smithy* shows an almost pastoral scene, as a blacksmith shoes a horse surrounded

Reproductions of Incidents of the Boer War.

————→•←————

(Arranged under the supervision of an experienced military officer from the front).

Shooting the Spy.

SCENE outside a guard room, with sentry on duty. An escort comes up with captured Boer spy, who is eventually shot.

Code word—**Spy.** Length **60** feet. Price **45s.**

Bombardment of Mafeking.

THE British soldiers are sitting round the camp fire. Several shells explode near them, causing much amusement.

Code word—**Mafeking.** Length **60** feet. Price **45s.**

A Camp Smithy.

SPLENDID scene of the camp smithy, with horses being shod, &c.

Code word—**Farrier.** Length **60** feet. Price **45s.**

Attack on a Picquet.

A BRITISH outpost is seen gathered round a camp fire, when a party of Boers steal out from an ambush, club their sentry and fire on the soldiers from all sides.

Code word—**Picquet.** Length **40** feet. Price **30s.**

Wrecking an Armoured Train.

A GRAPHIC and complete reproduction of the armoured train incident at Mafeking. The British are seen defending the train and firing on the Boers. Several are wounded, and at last the British officer hoists a white flag in token of surrender.

Code word—**Train** Length **100** feet. Price **75s.**

Nurses on the Battlefield.

A MOST affecting picture, but very beautiful and natural. It depicts the battlefield with the wounded and dead scattered over it. The picture shows the stretcher party with doctor and his orderly, who, with the nurses, are tending a wounded Boer. At the same time a British soldier is carried down by his comrades to the other nurses. *Specially recommended.*

Code word—**Nurses.** Length **60** feet. Price **45s.**

48. Page from Paul's 1900 catalogue advertising "reproductions" of scenes from the war, as staged on Muswell Hill golf course.

by resting soldiers, while *Nurses on the Battlefield* shows both Boer and British casualties being looked after.

It seems appropriate that the first of these reproductions should have been found in New Zealand, since this was the first colony to volunteer troops for the conflict.[34] Not only in Britain, where so many families had a personal stake through husbands, brothers, and sons serving, but throughout the empire, there was intense interest in the war's progress. This seems to have been satisfied in part by screening Paul's reproductions along with true actualities from the Cape, which were being supplied in increasing numbers by Warwick and Biograph. Paul's smaller output of on-location films focused on technical and logistical subjects, such as a giant Royal Engineers' observation balloon or *Telegraphing Casualties*, showing "the first time the system has been used in war"—a subject bound to interest an electrical engineer. Meanwhile, other staged films purporting to represent scenes from the war were being supplied by Mitchell & Kenyon and Edison, with fewer scruples about their claimed authenticity.[35]

It is only in rare cases that we can discover in what combinations these "war films"—none lasting more than ninety seconds—were shown. A program of "patriotic pictures" at the Leeds Empire music hall in January 1900 was reported to have "stirred deeply the emotions of the crowded audience."[36] In July, a program of "4000 feet of living pictures," containing "all the latest films of the Transvaal War," was widely advertised in New Zealand.[37] This would have run for over an hour and would almost certainly have included both staged films and actualities, no doubt from different sources. But one occasion, in Melbourne on 10 February, was reviewed in enough detail to identify three of Paul's reproductions among the "half a dozen of cinematograph films" that were interspersed with lantern slides of "leading figures, both Boer and British."[38] The reviewer admitted to being puzzled by "how a cinematograph operator could possibly be a witness of such episodes," apparently accepting them as authentic, but recorded the "sombre fascination" the show had for an audience that probably included relatives of Australia's "Bushmen" volunteers.

Allegory and Affect

Among these increasingly common genres of war film, Paul produced two very different series that stand apart from the output of his competitors. The first, which might be deemed propaganda, were advertised by Paul as "New Patriotic Trick Films," referring to their use of stop-motion and other optical effects. These were not the first overtly patriotic films: Paul had produced a close-up of the Union Jack in 1897, possibly to include in Diamond Jubilee programs, and Edison had released a variety of US flag images during the Spanish-American War of 1898. But the four patriotic films Paul made during 1900–1901—*Kruger's Dream of Empire, His Mother's Portrait; or, The Soldier's Vision, Britain's Welcome to Her Sons*, and

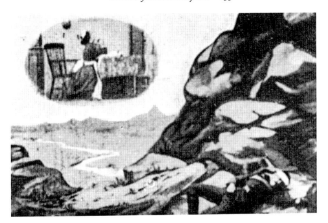

49. Catalogue illustration of *His Mother's Portrait; or, The Soldier's Vision* (1900), in which Paul "remediated" a popular poetic and theatrical effect, as a wounded soldier dreams of his mother's gift.

Britain's Tribute to Her Sons—are considerably more elaborate, offering highly compressed narratives and allegorical tableaux.[39] *Kruger's Dream* (May 1900), drawing on the long English tradition of political caricature, showed the Boer leader dreaming that he might be offered the English crown, only to awake and read a reminder that his general Cronjé has surrendered, before he too is captured by a squad of British soldiers and transformed into the figure of Britannia, with the soldiers forming "a tableau representing 'The Defence of the Empire.'" Like the transformation films of Méliès and Gaston Velle, the sequence made use of the newly discovered potential of stop-motion for "magical effect," thus renovating the patriotic stage tableau as Kruger's supposed ambitions are mocked and frustrated.

His Mother's Portrait, subtitled *The Soldier's Vision* (July 1900), invoked the "vision scene," a device that had been central to nineteenth-century theater and illustration: indeed, the film's "vision" amounts to a compressed melodrama that could well have played out on the Victorian stage.[40] A CIV volunteer takes leave of his aged mother, who gives him her portrait in a gilt frame. Later, lying wounded on the veldt, he dreams of his mother, "and slowly the vision appears in the sky of the room in which he last saw her," where she is reading his last letter. In the theater, this vision would have been achieved by raising the lights on another scene behind a gauze scrim, but film framed it as a fleeting vignette superimposed above the foreground image of the soldier—a striking example of how early filmmakers like Paul and Méliès were inventively adapting familiar effects, which audiences would easily grasp, to the new medium.[41] Finally, a nurse and a surgeon find the soldier, "as if in answer to her prayer," and discover that his life has been saved by the portrait he was carrying. Here, we might say, the maternal nation extends her protection to her distant sons.

Allegory was also the basis of Paul's most overtly imperialist war films: *Britain's Welcome to Her Sons* (September 1900) and, a year later, *Britain's Tribute to Her Sons*. In both, the traditional warrior-mother figure of Britannia, long familiar from coinage and statuary, is central.[42] The dramatic ancestry of this image lies in earlier forms and media, starting with the seventeenth-century masque;[43] continuing with patriotic tableaux in such popular venues as Astley's Ampitheatre, as well as music halls;[44] in the posed "living pictures" that were a staple of amateur theatricals; and in lantern slides and the host of cheap printed media that by now included stamps, postcards, cigarette cards, and many other such tokens.[45] Paul and his collaborators, who now included the conjuror Walter Booth and ex-music-hall artist Lewin Fitzhamon, were thus bringing into film an intermedial image that had often been mobilized in times of war.[46]

The two films extended this already rich tradition. In *Britain's Welcome*, Britannia draws back a curtain to reveal a "rustic cottage," where a one-armed pensioner awaits the return of his son, who receives a Victoria Cross from Britannia before the heroic action that earned it is also seen. The scene then "fades to the home of an officer, where a wife and child are reading a letter as the returning husband appears before … he is seen leading his men in attack." In a final transformation, Britannia appears on the seashore "in front of a large group of national flags, with supporting sailors, while the two family groups appear and the soldiers grasp hands, with Britannia adding her hand," thus, as the catalogue explains, "ratifying the bond of fellowship and the levelling of class and the common cause of Queen and Country."[47] This film of less than two minutes uses an eclectic visual rhetoric that is both archaic and modern to link the stock music-hall figures of Britannia and the veteran with the themes of class unity and imperial solidarity, much as Kipling had done in "The Absent-Minded Beggar."[48]

The theme of *Britain's Tribute to Her Sons* is the comfort provided by "the rallying of the sons of Empire who have come to the assistance of the Mother Country in the Transvaal War." By September 1901, when the film appeared, there had been every appearance of imperial solidarity, and it featured "representatives of various Colonial regiments." Perhaps sensing that something more than a mere tableau was needed, Paul also offered the film as one of a group of "songs with animated illustration," with music available for hire.[49] Sets of slides illustrating popular songs had been popular for decades, and there were already many devices on the market for synchronizing prerecorded music. But Paul's "animated illustrations" seem to have been intended for inclusion in a larger entertainment that would have other musical elements—perhaps in a church hall, like the "Trip to the Transvaal" lantern lecture, or at a theater such as the Royal Strand, where the song's composer, Frank Byng, worked.

Whatever their motivation, the patriotic trick films seem to have proved short-lived attractions, or perhaps were simply commercial failures, disappearing from Paul's catalogue by the 1902 edition. This may equally reflect the growing unpopularity of the South African war, as it dragged on through 1901, with British and imperial forces largely in control of the Boer republics but still vulnerable to their opponents' guerrilla tactics.

The First Documentary Series?

Meanwhile, Paul had embarked on another ambitious strand of production. His series *Army Life*, subtitled *How Soldiers Are Made*, covered a remarkable range of activities in thirty-three short films, ranging from 30 to 160 feet, and was intended to be shown as a dedicated program. The series was clearly prompted by the war, and his brothers being volunteers may have had a bearing, but his earlier experience with series, such as the tours of Spain and Portugal, Egypt, and Sweden, also undoubtedly contributed to the idea. Apparently filmed by Paul himself over the summer of 1900, with a total running time of seventy-five minutes, this would remain his single most ambitious film enterprise.[50] Starting with "Joining the Army," the series works through the routines of depot and camp life, basic training, and gymnastics, before showing the training of cavalry, artillery, and engineers in considerable detail. One press review noted that "the military authorities have taken a deep interest in Mr Paul's work, and have afforded him unwonted facilities."[51] The series' link with recruitment was acknowledged in Paul's foreword to the lavish illustrated brochure he produced, as well as in a central section portraying "a large body of Reserves called up for service in South Africa [showing] the pleased expression on the faces of the men at the prospect of active service again."[52]

Paul proudly proclaimed *Army Life* as "under the patronage of the Commander-in-Chief," namely the army's adjutant-general, Sir Evelyn Wood, and premiered the program at a special afternoon screening at the Alhambra on 18 September 1900. From what must have been an impressive, and gratifying, occasion he was able to print enthusiastic press coverage quoted from no fewer than sixteen newspapers:

Series III.—CAMP LIFE AT
ALDERSHOT.

Firing at the Ranges.

THE picture is in two sections, the first showing infantry firing volleys in the various positions. In the second part the method followed by cavalry, firing carbines, is clearly shown.

Code word—**Ranges.** Length **60** feet. Price **60s.**

The Soldiers' Bathing Party.

HERE the men are depicted coming down to the water-side with towels, and eager for their cooling swim. Scores of them jump in rapid succession from the spring board, diving into the water and swimming about in different styles, with plenty of fun and splashing.

Code word—**Bathing.** Length **60** feet. Price **60s.**

50. *Army Life* (1900). A page from the brochure advertising Paul's comprehensive suite of films on training in all branches of the British Army, evidently produced to stimulate recruitment during the Anglo-Boer War.

A large company, consisting mainly of Army officers and their friends, assembled at the Alhambra Theatre yesterday afternoon to witness a reproduction upon a screen of animated photographs, taken by Mr R. W. Paul MIEE. . . . This preliminary exhibition of the photographs was extremely satisfactory. (*Times*)

Altogether in an hour and a quarter the spectator sees more of the life of the Army than he would witness in months of travel, and the exhibition promises to be popular, as was shown by the outspoken approval of a company including many hundreds of officers and men of all branches of the service. (*Daily News*)

There is an intimacy about these peeps behind the scenes that is now always found in such pictures, and that the views were true to life was attested by the attitude and comments of the Duke of York's boys regarding the earlier scenes, and of the Chelsea pensioners regards the later. There was perfect recognition in the laughter of the boys at the blunders of the young recruits, and unmistakable reminiscence in the chuckle of the veterans over the struggle at the door of the newly-opened canteen. (*Morning Post*)

It is doubtful whether the Animatograph has ever been turned to such good account. . . . Of exceptional merit, combining art and actuality. (*Echo*)

Without doubt, a veritable triumph over all systems of animated photography which have, up to the moment, been presented in London. (*Observer*)

The films were interspersed with printed slides, containing "a quantity of useful and interesting information"—recommended as "essential where a lecturer is not employed." After the Alhambra premiere, the *Army Life* series was shown nightly in October at the Royal Agricultural Hall in Islington, while potential exhibitors around the country were invited to offer dates.

We do not know what scale of response this unusual offer produced. In his 1901 catalogue, Paul quoted one satisfied exhibitor, the music director of the Drury Lane Theatre (who was also organizing programs for the newly enlarged Kursaal pavilion in Bexhill on Sea), asking if he might arrange "a sort of permanent hire system." Meanwhile Paul assured "responsible exhibitors" that renting the films would be "a very profitable transaction," and offered as well a selection of "suitable music" to accompany the films, arranged by Frank Byng of the Strand Theatre.

By the time peace was finally agreed between Britain and the Boer Republics on 31 May 1902, after months of stalemate, the war films had largely disappeared from Paul's catalogues, and, as will be seen in the next chapter, he was busy experimenting with new and ambitious forms of drama and fantasy. At the very least, however, as John Barnes claimed, the war had given "cinema the fillip it needed," prompting experiments with new genres and modes of relation to audiences. It had also kick-started a rental market for film, as Richard Brown argues, due to exhibitors' unwillingness to invest heavily in subjects that might prove of only temporary interest.[53] Since this would become the commercial basis of all film distribution and exhibition a decade later, it may indeed be the most far-reaching legacy of the war cinema.

Hedging British Bets

One other curiosity appeared among Paul's high volume of production in 1900, referring to another foreign war. *Yellow Peril* has sometimes been confused with a different film, *Extraordinary Chinese Magic*, but is revealed in a catalogue synopsis to be an elaborate allegory of the Boxer Rebellion that was currently convulsing China. Appearing in Paul's catalogue as an "original Patriotic Trick film," it showed a European conjuror in front of a Chinese temple, who plucks from the air "samples of one kind of yellow peril—i.e. gold." These accumulate in bags "labelled with the coin mark of one of the Allied Forces," until "the head of a Chinaman" appears, followed by a Boxer, who chases the magician and proceeds to devour the gold. The conjuror then reappears and severs the giant head, which opens to reveal "imps of Anarchy, Riot and Disorder" that are only quelled by "representatives of the Allied Powers," who transform them into "China's floral emblem, the Sunflower." In one final twist to this elaborate performance, Russia "goes to cut this gigantic plant from its stem" but fails, whereupon "Peace, waving her olive branch, emerges [and] the Powers lay down their flags at her feet [and] congratulate themselves on the happy termination of events."

What became known as the Boxer Rebellion, between August 1900 and the following September, was an uprising by the Yihequan, or Society of Righteous and Harmonious Fists, whose members blamed foreigners, and especially Christians, for China's ills, and massacred large numbers of missionaries and Chinese Christians. These killings, however, were soon overshadowed by the barbarity of the fifty thousand troops sent into China by an eight-power alliance in October 1900 to lift the Boxer-imposed siege of foreign legations in Beijing. Germany's Kaiser Wilhelm called openly for savagery by German troops, while both Russia and Japan tried to benefit territorially from the terms that were imposed on the dowager empress Cixi. Paul's film seems to reflect a high degree of understanding of the complex situation in China—much greater than the best-known surviving British film dealing with the rebellion, James Williamson's *Attack on a China Mission*, released in November 1900. Filmed

at a villa near Brighton, this purports to show a missionary family surrounded by attacking Chinese, who are dramatically rescued by a detachment of bluejackets, or marines, revealed advancing in what is considered
one of the earliest "reverse shots" in filmmaking.[54]

Paul's "patriotic trick films" relating to the South African and Chinese
wars seem overelaborate for the contexts in which his films were generally shown, so may have been intended for specialized audiences, ideally
with some form of commentary. This is not impossible, although no record has survived of such occasions, any more than there is evidence of the
actual take-up for his *Army Life* documentary series. There is a rare report
from 1901, in an article by the noted magic lantern lecturer T. C. Hepworth (father of the filmmaker Cecil), of an enthusiastic London showman setting off blank cartridges in his theater to enhance his audience's
experience of war films with the smell of gunfire.[55] The Blackburn-based
firm of Mitchell & Kenyon, specialists in "factory-gate" films aimed at
local audiences keen to see themselves on screen, turned enthusiastically
to producing subjects, most of which were frankly propagandist, about
the Boer War (with titles like *White Flag Treachery* and *The Sneaky Boer*)
and the Boxer Rebellion (*The Assassination of a British Sentry*). Compared
to these, Paul's "Reproductions of Incidents of the Boer War" may have
been relatively restrained.

Four years later, British filmmakers responded to another distant war,
albeit one in which Britain had no obvious national interest. In February
1904, news spread of a surprise attack by the Japanese navy on the Russian fleet, moored at its base in Manchuria. This was soon followed by
Japan's declaration of war on Russia, triggered by a disagreement over
dividing Manchuria and Korea into separate spheres of influence. On a
diplomatic level, Britain was bound by a 1902 treaty to offer aid if Japan
was attacked by two or more countries, although not if it was merely one
opponent, as on this occasion.

Although British policy was to remain uninvolved in the conflict, there
were a variety of cultural reasons why many in Britain may have been in-
clined to support Japan over Russia, and particularly why Paul may have
been partisan. Fifty years earlier, the Crimean War had pitted Britain
and its allies against the Russian Empire, leaving a legacy in the many
place-names commemorating its battles. More recently, after the Meiji
Restoration of 1868, Japan had started a process of rapid modernization,
which involved building extensive relationships with Western countries,
including Britain. For example, the majority of ships in the Japanese navy
had been built in British shipyards, and in the 1870s the English scientist
William Ayrton had established a department of electrical engineering at

what would become Tokyo University. Ayrton, it will be recalled, was one of Paul's mentors at Finsbury College, and several of Paul's instruments were based on Ayrton's designs and concepts. Whether or not Ayrton had communicated an enthusiasm for modern Japan to his pupil, there was undoubtedly a sense that Japan, as another island nation, was now emulating Britain in its technological and industrial drive. One of Paul's releases in March 1904, *Naval Manœuvres*, showed HMS *Trafalgar* together with the Japanese ship *Niobe*, while in July 1905 he would film the launch of the last Japanese battleship built in Britain, *Katori*.

Meanwhile, on 21 October 1904, the Russo-Japanese War came bizarrely close to Britain. In a climate of high tension and rumors of enemy stealth, part of the Russian Baltic Fleet was sailing through the North Sea, on its way to the Pacific Ocean. One Russian ship mistook a group of Hull trawlers for Japanese torpedo boats and signaled for others to open fire. One trawler was sunk, with two crew members killed, and four others damaged, with further injuries. In the confusion, Russian ships also fired on members of their own squadron, damaging the cruiser *Aurora*, which would later play a leading symbolic part in the October Revolution in Petrograd.[56] What became known as the "Dogger Bank incident" was popularly referred to as the "Russian outrage," and on 29 October Paul released *Russian Outrage in the North Sea*—a five-film series matching the title of a simultaneous Warwick release. Since the Paul films were reissues of previous fishing films, it seems likely this was opportunistic "rebranding" to capitalize on the incident.[57]

More significant are the three films Paul released in February–April 1905, each dramatizing some aspects of the land war. *An Affair of Outposts*, advertised as "a thrilling reproduction of an Incident of the War in the Far East," showed a Russian scouting party being ambushed by Japanese troops, who then use a captured plan to successfully attack a Russian camp. After the Russian commander hands over his sword in surrender, "his flag is lowered and the Japanese flag raised in its place." Paul's catalogue extolled the film's authenticity—"no expense has been spared to make it correct in every detail of dress and action"—and promised "wild and picturesque scenery," with actors "selected as representative of the respective nationalities." In *All for the Love of a Geisha*, a boy sacrifices himself so that an Englishman and a Japanese can elude Russian soldiers and rescue a geisha who is being threatened with forced marriage by them. When they return the geisha to her teahouse, the Englishman embraces her and "shakes hands with his brave comrades," before "two Japs hold above them the crossed flags of England and Japan, making a tableau, typifying the Anglo-Japanese alliance." *The Capture and Execu-*

April, 1904.

PAUL'S FILMS.

Uniform Price

per 6^D foot.

ROBT. W. PAUL, *Animatograph Depôt,*

68, High Holborn, LONDON, W.C.

Telegrams : "Calibrate, London." Telephone : 4814, Holborn.

AN AFFAIR OF OUTPOSTS.

A Thrilling Reproduction of an Incident of the War in the Far East.

51. *An Affair of Outposts* (1904). The Russo-Japanese War of 1904 prompted a number of British films, mostly pro-Japan, including Paul's drama, which boasted of care taken with authentic uniforms and settings.

tion as *Spies of Two Japanese Officers* apparently showed its protagonists posing as "coolies" to blow up a train, before being discovered and paying with their lives.

It is conceivable that the geisha story was influenced by Puccini's latest opera, *Madama Butterfly*, which was being widely performed during 1904, or by the play on which it was based, which the composer had seen in London in 1900. In any case, both it and *An Affair of Outposts* showed Paul scaling up two of the genres he had first produced for the Boer War: realistic battlefield dramas and elaborately symbolic tableaux. More conventionally symbolic were *Jap Versus Russian* (March 1904), in the form of a wrestling match, and *A Russian Surprise* (April 1904), in which an Asian waiter serves an exploding dish to his visibly Russian guest.

Paul was not alone in producing pro-Japanese films at this time. His ex-employee Frank Mottershaw, now back in Sheffield, made two "reproductions" of military action that featured Japanese protagonists defeating or escaping from Russian troops. More evenhandedly, Charles Urban released a steady stream of "topicals" covering both Russian and Japanese military actions, as well as everyday life in both countries.

There would be a further echo of the Russo-Japanese War from Paul in 1905, as indeed there was in Russia itself. During the second year of the

52. *Goaded to Anarchy* (1905). The revolution that swept Russia in the aftermath of the war with Japan was marked by what would be Paul's most elaborate costume drama. Though lost, its plot and stills are available from a catalogue, condensing the familiar narrative of tsarist repression and violent anarchist retaliation, followed by exile to Siberia.

war, with Japan repeatedly victorious at sea and on land, discontent escalated in Russia. A massacre of demonstrators in St. Petersburg in January, remembered as Bloody Sunday, marked the beginning of a wave of protests, strikes, and repression which would later be considered the "first" Russian Revolution, anticipating the two in 1917, which culminated in the Bolshevik seizure of power. One feature of this period was an increased level of terrorist activity, with bombs and assassinations becoming commonplace. Paul's film *Goaded to Anarchy*, released in September, offered a condensed account of events that would have seemed familiar to many. During a police raid on a secret society meeting, a policeman is killed and a girl arrested for having an incriminating pamphlet. The commanding general refuses clemency, and by the third scene she is part of a group of convicts "getting ready to march to Siberia." As the convicts "wearily trudge through the snow," they are "treated more like dogs than human beings." Meanwhile, a "young patriot" is seen "experimenting with chemicals of an explosive nature" to create a bomb, which explodes in the saloon of a palace, killing the general. According to the detailed catalogue description, "revenge has slowly and surely overtaken the man who showed no leniency to those at his mercy."

Tsar Alexander II had in fact been assassinated by bomb-throwing anarchists in 1881, which led to an intensified repression of radicals by his successors. But political violence was not confined to the Russian Empire. During the last decades of the previous century, London had experienced a number of politically motivated bombings, carried out by Irish republicans and a French anarchist. The suppression of an uprising in Latvia in 1905 led to an influx of refugees reaching Britain, which would produce three violent "outrages" in London during the years 1909–1911, culminating in the "Siege of Sidney Street." All of which may help explain the topicality of Paul's unusually elaborate and forthrightly political production.

Nearly forty years later, Paul recalled "the demand for something more exciting" than could be captured on location in South Africa, acknowledging that the showmen would present his and other films as dramatically as they could. But his Russo-Japanese War dramas of 1904, followed by the portrayal of Russian repression in *Goaded to Anarchy*, showed him aspiring to an ambitious level of dramatic realism, considerably before this trend began in French production and in American Vitagraph around 1908.[58]

CHAPTER 8

Telling Tales

STUDIO-BASED PRODUCTION

By 1898 "animated photography" was no longer a novelty. Paul later re-called that "in the year after the Jubilee, the public interest in animated pictures seemed to be on the wane," despite his company's efficiency in supplying topical subjects promptly.[1] There were more and more com-petitors offering projectors, cameras, and ever-increasing numbers of films — even though Paul remained the most prolific British producer.[2] To maintain this position, along with his equipment and instrument busi-nesses, Paul took the bold step of buying a four-acre site in the developing North London suburb of Muswell Hill, on which he would build a studio complex "to secure space for taking subjects on a more ambitious scale."

In doing so, he may well have been aware of similar moves across the Channel. There, two of his former customers, Georges Méliès and Charles Pathé, had built studios, the former a simple glass-house struc-ture in the Paris suburb of Montreuil at the end of 1896, and the latter a studio and factory in the riverside town of Chatou, ten kilometers west of Paris, started at the same time, in 1898.[3] Like Pathé and the other emerg-ing French film industrialist, Leon Gaumont, Paul was offering a wide range of equipment as well as films, and he clearly saw the advantages of moving manufacture out of the city.

What motivated Paul was also, and perhaps more importantly, a vision of the potential to make "long films giving complete stories." By October 1898, he was ready to announce a major initiative in an advertisement that amounted to a manifesto:[4]

The public have been surfeited with Trains, Trams and 'Buses, and be-yond a few scenes whose humour is too French in nature to please English audiences, the capacity of animated pictures for producing BREATHLESS SENSATION, LAUGHTER AND TEARS has hardly been realised. The DAY IS PAST when anything in the way of animated pictures will do for an

A FULLY EQUIPPED STUDIO, with Scenery and Accessories, for the production of set Scenes, is available at my Muswell Hill Works for the use of Customers by special arrangement.

53. *Top*: Paul's filming stage at Sydney Road. Early in 1898, Paul bought land on the outskirts of North London, in what is now Muswell Hill, to build Britain's first studio complex. This illustration from his 1900–1901 catalogue shows the stage, later described as "The Large Studio, with paint rooms for full-sized scenery . . . and a stock of about fifty backgrounds ready for use." Also "stage-traps, bridges and wings" enabling "any stage performance to be animatographed." *Bottom*: Newton Laboratories, shown in the 1901–1902 catalogue, "having a capacity for 1½ miles of finished films per day, with site for extension."

audience. Exhibitors and Managers have been asking for something New, Distinctive, Telling and Effective.

The passing mention of "too French" humor must refer to early films by Méliès and Gaumont that were breaking new ground, at least in subject matter.[5] (Paul would again invoke "English taste" in 1901 to promote *The Magic Sword*.) But by the autumn of 1898, after what must have been months of intensive preparation and investment, he was ready to announce a major new emphasis on high-quality fiction:

> A staff of Artists and Photographers have been at work in North London, with the object of producing a series of animated Photographs (Eighty in Number), each of which tells a tale, whether Comic, Pathetic or Dramatic … with such clearness, brilliancy and telling effect that the attention of the beholders should be riveted.

This ambitious new challenge to his competitors was lavishly advertised in the first of Paul's catalogues to be illustrated with frame enlargements—and with titles given in French and German as well as English, implying that he was either already selling to near European countries, or hoped soon to do so. Product differentiation was also emerging. The catalogue offered films from 80-foot "double length" at £3 to "special series of cheap films for amateur exhibitions" at £3 for four, with a further list of "Sporting, Domestic, Travel, Exercise and Jubilee Subjects."

Beyond promising higher production values, Paul recognized the importance of finding ways to extend the single-scene anecdotal structure of early filmmaking. He started with the relatively simple step of linking pairs of films that could be shown singly or consecutively. Thus *A Rescue from Drowning* was followed by *Rescued*, in which spectators cheer the rescue, while *The Sailor's Departure* was complemented by *The Sailor's Return* (production of which, as noted in chapter 5, may have led to the accidental death of the Pauls' son).[6] One film from this intensive period of production stands out, and is the only one that has at least partially survived.[7] *Come Along, Do!* took its subject and title from a comic situation had already been a subject for almost every nineteenth-century form of entertainment, from popular song and stage sketch to stereographs and lantern slides.[8] In the first shot, a "country couple" is seen outside what is labeled an "art exhibition," eating a meal from their basket "to the amusement of passers-by," as the catalogue description explains. The pair then walk toward the entrance, and in the second scene, set in the interior, the man is seen inspecting a nude statue of Venus "with some glee," before his

54. Catalogue still of the second scene from *Come Along, Do!* (1898).

wife angrily pulls him away "with a most amusing expression." The fact that Robert and Ellen Paul are believed to have played the couple suggests that they were conscious of taking an important step beyond what had hitherto been attempted in production.[9] Given her theatrical experience, it seems likely that Ellen would have played an important part in launching the studio, and proposing well-tried stage subjects such as this.

Not only did *Come Along, Do!* involve fictional characters and settings, but its construction in two separate tableaux makes it the earliest surviving fictional film to create narrative continuity between successive shots.[10] Indeed, this simple "outside and inside" drama concisely foreshadows the expanding future of narrative cinema, showing how time and space could be streamlined. The same period saw Paul advance boldly into this new narrative form, with *Our New General Servant* consisting of no fewer than four spatially distinct scenes: a woman hires a servant from an agency, with whom her husband then flirts in the parlor and garden, before the servant is dismissed. Although the film no longer survives, it appears to have featured another innovation with far-reaching implications: the inclusion of on-screen intertitles.

The construction of his ambitious new studio complex would provide opportunities for at least two dramatic subjects. *The Bricklayer and His Mate*, subtitled *A Jealous Man's Crime*, must have been filmed on the building site, as was *Thrilling Fight on a Scaffold*, in which a man appears to be injured in a fall. And once completed, the studio's stage machinery made possible more elaborate narratives. *The Miser's Doom* (1899) seems to have been Paul's first foray into the supernatural, using double exposure for the "apparition of a woman in a ragged dress . . . holding a child to her breast" who scares the miser to death, after which "the ghostly

55. *Upside Down* (1899) made use of Paul's new studio facilities, with a domestic backdrop that would recur in other films.

visitant slowly fades away." This film has not survived, but one of Paul's most surprising productions from the same period has, and shows him tapping another rich vein of topical humor, making unexpected use of a continuity edit. We might not guess that *Upside Down; or, the Human Flies* was a satire on the popularity of spiritualism without the catalogue's explaining that "a party of four are practising table turning, when a professor of spiritualism is shown into the room." This opening is missing from surviving prints, which start with the "professor" lifting the table to prove the absence of any trickery, before he makes one of the men's hats fly upward. Apparently this is done to explain that "he will invert the party in the same manner." He then disappears, and the four suddenly find themselves standing on the ceiling.

The trick was achieved by a cut to a second shot in which both the camera and the painted wall and furnishings were inverted, so that the décor appears unchanged while the performers are upside down. The men start boisterously tumbling playing leapfrog, with one of the women waving an open umbrella to emphasize the disorientation. Even after more than a century of increasingly sophisticated filmic tricks, it remains a striking effect, apparently achieved by the simplest means. It may also be evidence that Paul's production was benefiting from a new recruit, the conjuror Walter Booth. Modern listings have attributed many of Paul's films over the next four years to Booth as "director," and the fact that he can now be identified as the "professor of spiritualism" in *Upside Down* certainly shows he was involved.[11] However, to label him a director at this early stage of production seems premature, not to say anachronistic.

Filmmaking at New Southgate in 1899 is likely to have been a collegial business, even if the arrival of Booth may have introduced a new emphasis on subjects requiring a magician's talent for managing expectation, as well as optical and other skills.[12]

Spiritualism and the dubious practices of mediums were a consistently popular theme in Victorian and Edwardian Britain, reflected in such varied texts as Robert Browning's monologue "Mr Sludge, 'The Medium'" and George and Weedon Grossmiths's humorous novel *Diary of a Nobody*.[13] Revealing the tricks used by mediums had been a popular premise for showman magicians such as Nevil Maskelyne, in his long-running show at the Egyptian Hall, billed as "England's home of mystery."[14] And it would be the subject of Paul's later *The Medium Exposed*, also provocatively titled *Is Spiritualism a Fraud?* (1906).

Remediated Classics

The majority of early fictional film subjects were comic, often based on traditional comedy turns such as the cheeky servant or tipsy monks, but ambitious producers were beginning to look further afield, offering scenes from popular narratives. G. A. Smith led the way, also in 1898, with *Cinderella*, *Faust and Mephistopheles*, and a take on Alexandre Dumas's *The Corsican Brothers*, using the earliest known double-exposures for "supernatural" effects. Paul's contribution was to bring Britain's most popular modern author to the screen for the first time in Britain, with *Mr Bumble the Beadle*, as part of his prolific 1898 output.[15] In Charles Dickens's *Oliver Twist*, Mr. Bumble is both a sadistic and a pathetic character, who loses the authority that has allowed him to tyrannize Oliver and many other orphans after marrying Mrs. Corney. Paul's short film, now lost, apparently showed only Bumble's courtship, although it would lead to two longer Dickens "adaptations," *Mr Pickwick's Christmas at Wardle's* and his longest extant fictional film, *Scrooge; or, Marley's Ghost* (1901).

At 620 feet, *Scrooge* offered an extended, multiscene version of Dickens's *A Christmas Carol*, long a Victorian Christmas favorite.[16] The film's simplification of the original supernatural tale relied heavily on both J. C. Buckstone's popular stage adaptation and the audience's presumed familiarity with the story. Yet Paul—almost certainly helped by Booth—was able to portray some aspects of Ebeneezer Scrooge's redemptive vision in new ways, due to his growing mastery of "trick film" techniques. Scrooge is met at his own door by the accusing face of Marley's ghost, anticipating what awaits him as he falls asleep, when the curtains of his

SCENE IV.

The Christmas
that might be.

Marley's ghost shows Scrooge his own
Grave and the death of 'Tiny Tim.'

56. *Scrooge; or, Marley's Ghost* (1901). Paul's longest and most elaborate production to date survives at half its original 600-foot length. It is notable for its extensive use of superimposition for Scrooge's haunted visions, and for what may be the first appearance of printed titles on screen.

room become a screen on which scenes appear of missed opportunity in Christmases from his youth. Both the superimposition of Marley's head and the vertical wipe that marks Scrooge's entering his house were effectively innovations, going beyond the vignetting and stop-motion effects already used by Smith and Méliès in films with a single setting.[17] Here they were used transitively, to advance the narrative, as Scrooge enters the dreamworld that will lay out his life of missed opportunity.

Although it is customary to present the early years of filmmaking as "progress" toward developed forms of narration, which *Scrooge* was in terms of its deployment of filmic techniques, it is surely more accurate to identify Paul's single-scene Dickens films and his *Last Days of Pompeii* of 1900 as something other than "adaptations." Clearly, he made no claim that these films "contain" the complex narratives of the works whose titles they bear. Rather, they should be seen in the same light as paintings that portray crucial moments in a narrative—whether biblical or historical or drawn from contemporary life. The "added value" of filmic portrayal, even at two minutes duration, was the novelty of lifelike photographed movement within spaces that represented the fictional world of the original. Another way of understanding this is through what Jay David Bolter and Richard Grusin have called "remediation."[18] According to their analysis of digital media, "new visual media achieve their cultural significance by paying homage to, rivaling, and refashioning earlier media."

If this interpretation is projected back to the early years of "animated photography," we may assume that the viewers of these early tableaux were not expecting condensed versions of a Dickens novel or of Bulwer-Lytton's *Last Days of Pompeii*. Rather, they were being offered the novelty of seeing a token of those familiar and respected cultural works "remediated" in a novel form. In the case of *The Last Days of Pompeii*, the spur to presenting this as a single scene of the ancient city collapsing on its citizens may well have come from another of that novel's many adaptations, when it was presented near Paul's studio in North London as a "pyrodrama" at Alexandra Palace, using spectacular fireworks for the eruption of Vesuvius.[19]

Film could not yet aspire to rival spectacle on such a scale, but it could offer a sense of immediacy and even plausibility, as Paul and other pioneers were discovering. In his 1936 text, Paul chose to recall the contemporary response to two of his films from this period, *Diving for Treasure* and *A Railway Collision* (both 1900). Both were "faked" — one by shooting men dressed in diving suits through a tank containing small fish, the other by filming model trains — yet apparently convinced viewers of their reality. Paul remembered the Prince of Wales and Lord Rothschild seeing them at the Alhambra and wondering how he had filmed underwater, while the rail film "was considered very thrilling ... and had a large sale," as well as being extensively "pirated" in America. In each case, Paul discovered, "the scene appeared sufficiently natural" to its initial viewers, despite or because of their growing familiarity with filmed images on screen. In Bolter and Grusin's terms, this is the "transparency" effect of a new medium, which disappears for those experiencing its immediacy.

As the range of Paul's subjects expanded, together with the technical resources of his newly completed studio, an ambitious new genre of fantasy began to be developed. One of the earliest examples of a fantasy that does not rely on a preexisting text, such as *Scrooge*, appeared in the same month, November 1901. *The Magic Sword*, subtitled *A Mediaeval Mystery*, exploited the same skills of superimposing images of different scale, and to judge from the unusual space devoted to it in the 1901 catalogue must have been considered a real achievement:

A sumptuously produced extravaganza in three dissolving scenes, with many novel and beautiful trick effects, now introduced for the first time. The period of this dramatic mystery is in the middle ages, and the facts of the actors and costumes being Old English, together with the original nature of the plot, cannot fail to please English-speaking audiences, which have become weary of foreign pictures of this class.[20]

The tone of this implies a great deal of planning and a real sense of author-ship, quite apart from the bid to flatter "English-speaking audiences," de-spite its being a film with only an elaborate title card and no intertitles! The most obvious "foreign pictures" with period costumes and medieval trappings would be those of Méliès, whose recent *L'antre des esprits* (*The House of Mystery*), like *The Magic Sword*, featured a magician in a cave and a levitating woman. Paul seems here to be deliberately proposing a rival English "dramatic mystery," although without Méliès's broad humor or any obvious literary or traditional source. Who, we might wonder, con-ceived and executed this, if it was not Paul himself, very likely in collabo-ration with his wife, drawing on her Alhambra ballet experience?[21]

The opening scene, set "on the moonlit battlement of an ancient castle," is indeed more elaborate than anything Paul had previously at-tempted, and involves no fewer than four supernatural figures in rapid succession. First, a ghost threatens a knight and his ladylove; then a witch tries to steal the lady, unsuccessfully, before "an enormous ogre no less than 15 ft high" succeeds in carrying her off. But as the knight despairs, a good fairy appears and gives him a "flaming sword." In the second scene, set in a witch's cave, the witch transforms her captive into a second witch, to confuse the knight, before producing from her cauldron "many differ-ent examples of magic" to distract him. These include a winged cherub, a smoke-blowing head, a skull, and finally two robed ghosts. More confu-sion ensues as the witch becomes a beautiful girl; then the knight recog-nizes his lost love, and the good fairy turns the witch into a roll of carpet, on which the pair escape as an explosion destroys the cave.

In the final scene, a party including the girl's parents is shown ban-queting "in the grounds of an old castle, seen in the distance." The lovers complete their escape by landing among them, "finishing with a striking and artistic tableau, over which the good fairy is seen hovering." The most obvious question is this: could any viewer in 1901 have followed such a complex and highly compressed narrative? The story seems to be an original one, despite making use of stock characters and situations, and so this seems doubtful. Although the catalogue entry fairly radiates pride in the elaborate production, not having a literary basis shared by maker and audience—as in, for instance, *Alice in Wonderland*—must have been a problem. The 1901 catalogue included another elaborate trick film, *The Haunted Curiosity Shop*, which offered a rapid concatenation of Gothic motifs, as an old curio dealer battles three female apparitions—a woman in two independently mobile parts, an "ebony skinned negress," and an Egyptian mummy—with skulls, a gnome, and a knight for good measure. Here, Paul seems to have ventured into a Méliès-style explosion of diabolical figures and motifs, without the attempted pantomime structure of *The Magic Sword* or the familiar narrative of *A Christmas Carol*. But there would be few examples of this genre in Paul's later production, as he turned increasingly to realistic contemporary settings for both comedy and drama.[22]

Chasing

Chase films are often assumed to be the first purely filmic genre — a form that can be traced back to the earliest attempts to structure movement on screen by making visible an anecdotal motive, as in the Lumières' *The Gardener Watered*, when the victim of a boy's prank gives chase around the garden. At almost exactly the same moment, in early spring 1895, one of Paul and Acres's first batch of films, *Arrest of a Pickpocket*, showed a man running to escape from a policeman, only to collide with a sailor and be wrestled to the ground by the pair. Filming some kind of vigorous action where there is an incentive to escape might be considered an obvious trope, and even if James Williamson's *Stop Thief* (1901) is widely regarded as the "first" chase film, there would soon be many examples from other pioneer filmmakers. In early issues of his new trade journal in 1907, Theodore Brown published an American article that claimed the Mutoscope company had invented the chase subject, but chose to doubt this in his commentary:

> A "chase" seems such an obvious idea that we suspect it figured in some of the very earliest subjects put out.... Whoever invented it, every firm has made a most liberal use of it, perhaps too liberal. A large proportion of the subjects produced at the present time are of this variety.[23]

As we have seen, Paul appears to have been the first producer to link spatially distinct images with an implied continuity of action in 1898 (in *Come Along, Do!*). Eighteen months later, he took an important step toward the extended chase with another two-shot film, *The Hair-Breadth Escape of Jack Sheppard* (March 1900), based on the exploits of the celebrated eighteenth-century thief. According to the catalogue:

> Jack is, when the scene opens, making love to the inn-keeper's daughter, but an alarm is given by the Boniface, who has caught sight of the watchman coming. The girl shows a way of escape, and the scene changes to the roofs; the pair throw a plank across the street, and carefully make their way across. When the watchman and his assistant attempt to follow, Jack,

57. *The Hair-Breadth Escape of Jack Sheppard* (1900). Judging by the catalogue illustration, this lost film about a notorious eighteenth-century highwayman used an elaborately perspectival backdrop.

lifting one end of the plank, throws them one after the other, 30 feet into the street below.[24]

The catalogue image, which is all that survives of the film, shows an angled perspectival backdrop of the street, with a plank balanced across the foreground. Presumably the first shot had an interior backdrop, with Sheppard engaged in his equally notorious custom of seduction. Like another "historical" film made at the same time, *The Last Days of Pompeii*, it was only necessary to show the most famous moment of the subject, as historical painters had long done—volcanic destruction, or bold escape—but now with the added frisson of movement within a tableau and the implied depth that lateral movement could provide.[25]

After these early "escape" subjects, it was one of Paul's first employees, Frank Mottershaw, who would take a decisive step toward the familiar multishot chase film. Mottershaw's *Daring Daylight Burglary* (1903), made in Sheffield after he had left Paul's company, tracked the police pursuit of a violent robber across a number of locations, ending with his arrest when leaving a train. Many copies were sold, and the film is credited with inspiring Edwin Porter's seminal *Great Train Robbery*, made later in the same year. Perhaps relevantly, another of Mottershaw's 1903 productions, *Robbery of the Mail Coach*, explicitly claimed to show "Jack Sheppard" as a highwayman.

The heyday of the European multishot, predominantly comic chase film would come around 1905, and in fact Paul's surviving *The Unfortunate Policeman* (also known as *A Victim of Misfortune*), released that year, was

58. *Top*: *The Last Days of Pompeii*. A catalogue still is all that survives of one of Paul's 1898 films invoking a popular literary classic: the volcanic climax of Bulwer-Lytton's novel was represented by a ceiling collapsing on the costumed actors with, apparently, Vesuvius seen in the distance. *Bottom*: A fireworks display, or "pyrodrama," of *The Last Days of Pompeii* was a popular attraction at Alexandra Palace, near Paul's studio, and may have inspired his film.

59. *A Victim of Misfortune*, also known as *The Unfortunate Policeman* (1905). Escalating comedy chases were becoming a staple of early cinema in France and America when this version appeared. Starting against a studio backdrop, with paint tipped over a policeman as punishment for flirting with a servant, it rapidly moves into the streets around Paul's studio, as he and a growing crowd give chase.

code-named simply "Chase."[26] It opens with a studio set representing a shop front, with two painters at work, who are served tea by a servant. A policeman stops and kisses the servant, leading one of the painters to retaliate by pouring paint over him. The ensuing chase runs through the streets of Muswell Hill, with the policemen in pursuit of the painter, followed by an increasing number of offended parties. Finally, the painter jumps into a horse-drawn cab, pushes the cabman out and drives off, leaving the frustrated pursuers to berate the overheated policeman. Although Ferdinand Zecca's *La course des sergents de ville* (*The Policemen's Little Run*, 1907) is often considered a classic of the *course comique* chase, acknowledged as an influence on Mack Sennett's Keystone Kops, Paul's *Victim* could claim precedence in this international genre (and in Sadoul's view, may have been copied by Zecca).

The following year would see another vigorous excursion into the streets surrounding Paul's studio. *The Medium Exposed* falls into two distinct parts. In the first, a group gathers for a séance, after we have seen the medium briefing an accomplice with a long white wig who hides in a trunk. The medium insists on being tied to his chair to avoid any suspicion of fraud, and once the lights go out, a dramatic display of disembodied bobbing heads takes place. But when one of the guests turns on the light, the deception is revealed. The three male guests bundle the medium, still tied to his chair, into the same chest that concealed the accomplice; carry the chest downstairs and into the street; and commandeer a delivery man's handcart to send the chest down a steep incline.

60. *The Medium Exposed* (1906), also known as *Is Spiritualism a Fraud?*, starts with an elaborate dramatization showing how séances could be faked, leading to a variation on the chase, with the fraudulent medium paraded through Muswell Hill on a cart.

When the medium falls out, a policemen and a passerby place him upright on the cart, and the final shot shows him being paraded through a busy shopping street in New Southgate, like some latter-day witch on an improvised tumbril.

Poetics of Motoring

If *The Medium Exposed* follows *A Victim of Misfortune* in using the mechanics of a chase comedy to humiliate its protagonist, another film from the same year plays an even more fantastic variation on the chase format. Robert Paul was an enthusiastic pioneer motorist, and one of his early films has the intriguing title *On a Runaway Motor Car through Piccadilly Circus* (1899). Apparently this was a version of the popular "phantom ride," normally photographed from the front of a train. In this case the camera was undercranked—exposing fewer frames per second than usual—to create a sense of high-speed danger in central London when projected. Danger was in fact the most common motif in early motoring films, with two 1900 examples by Cecil Hepworth bearing equally threatening titles: *How It Feels to Be Run Over* and *Explosion of a Motor Car*. Paul's *The ? Motorist* (1906), by contrast, feels like a motoring fantasia with the underlying theme that motorists are the natural enemy of interfering officialdom.

The truncated version that has survived begins with a motorist and his lady passenger being signaled to stop by a policeman. When they refuse,

61. *The ? Motorist* (1906). A tour de force of stop-action editing and model-work, this fantasy of an elusive motorist escaping the law featured Paul's own car, in which he was fined for breaking the restrictive speed limit of the era in 1907. Both film and catalogue synopsis express many early motorists' feeling of persecution.

the policeman jumps onto the bonnet, and in the next shot is seen falling under the car's wheels — which does not prevent him resuming the chase. The car then drives directly at a building which it mounts vertically, to the amazement of onlookers, before driving over rooftops to reach outer space. Once round the disk of the moon, then a comet-assisted landing on the rings of Saturn. Falling off Saturn's rings, it plummets back to earth, crashing through the roof of a building that is revealed to be "Handover Court" in session. A policeman giving evidence against a motorist in the dock is interrupted as the car lands and drives off, chased by the policeman, judge, and others. The scene changes to a suburban road, where the original motorist and his passenger come to a halt, and the driver gets out to crank the car. The policeman springs out to catch him, but the car turns into a horse and cart, with driver and passenger now dressed accordingly, to the bafflement of the policeman. The driver climbs aboard, and as he sets off, the cart abruptly changes back into a car, and its occupants wave triumphantly at the bewildered policeman and his helpers.

The idea of a shape-shifting vehicle that can elude all pursuit unavoidably evokes for us later fantasy vehicles, from Chitty Chitty Bang Bang to the DeLorean used in *Back to the Future*. But in 1906, the setting of *The ? Motorist* is recognizably New Southgate: the early scene is set outside the Colney Hatch asylum, and the building that launches the motorist on his cosmic journey is the Orange Tree pub, while the car used was Paul's own

1903 Gladiator. This is confirmed by the report of a prosecution brought against Paul in the following year, for speeding at thirty miles per hour, in which he was found guilty and fined.[27] The prosecution, and the film that anticipated it, refer to what was in the early 1900s an ongoing dispute between British motorists and the authorities over what should be considered a reasonable speed limit, and especially the attitude of police forces, accustomed to setting speed traps on roads where motorists might be expected to exceed the limit of twenty miles per hour. Pioneer motorists, such as Paul and the author Arthur Conan Doyle, felt police were ambushing motorists, with over a thousand speeding fines imposed during 1904–1905.[28] This information was released to coincide with a Royal Commission report in July 1906, shortly before the film's release, which may have helped motivate its production. We can hardly discount Paul's own input, bearing in mind the strength of feeling he revealed in court a few months later.[29] Having been to a motoring club meeting, "before starting back he was told the police were timing cars." He asked a friend to "time the car from milestone to milestone," and so was able to reject the police evidence as "woolly," claiming the car "had never done more than twenty-four miles per hour in its life," but to no avail.

While *The ? Motorist* is inventively farcical and, like most of Paul's comedies, robustly anti-authority, it clearly registers a plea for motorists' rights. Its immediate precursor would be Méliès's *Paris to Monte Carlo in Two Hours*, released in the previous year, in which an obviously fake car creates chaos as it speeds from Paris to the Riviera.[30] Some sequences in Méliès's film use the same skeletal car image as does Paul in his aerial sequences, suggesting a direct influence. But closer to home, Paul's neighbor in Muswell Hill, the poet W. E. Henley, had published his *Song of Speed* in 1903, probably the earliest celebration of the exhilaration of motoring:

> This marvellous Mercédes,
> This triumphing contrivance,
> Comes to make other
> Man's life than she found it:
> The Earth for her tyres
> As the Sea for his keels;
> Alike in the old lands,
> Enseamed with the wheel-ways
> Of thousands of dusty
> And dim generations,
> And in the new countries,

Whose Winds blow unbreathed,
And their Lights are first-hand
From our Father, the Sun.

Henley's ode to the "marvellous Mercédes" gave motoring an epic and even mythological dimension, while another recent convert, Rudyard Kipling, would publish his *Muse among the Motors* in 1904, which included Byron falling victim to a speed trap.[31] In the following years, there would be growing variety of cultural references to motoring. For instance, Filippo Marinetti's *Futurist Manifesto* (1909) hymned

> a new beauty: the beauty of speed. A racing automobile with its bonnet adorned with great tubes like serpents with explosive breath … a roaring motor car which seems to run on machine-gun fire, is more beautiful than the Victory of Samothrace.[32]

A year earlier, Toad of Toad Hall had become the first reckless driver in literature, in what would become one of the most widely appreciated English novels for children, *The Wind in the Willows*. Unable to control his passion for speed, Toad crashes cars and endangers life, before he belatedly sees the error of his ways. Paul's mysterious motorist seems to combine elements of Henley's ecstasy and Toad's recklessness with his creator's own passion for automobiles, adding a magical ability to morph into a horse-wagon at moments when his freedom is threatened. From a contemporary response, we know that the film struck a chord with at least one distant viewer. The Russian symbolist Andrei Bely wrote in his 1907 essay "The City":

> A motor car, a bicycle, a policemen are flying in pursuit. Crump! The car smashes through a wall. … It's funny, but it's really not at all funny. Walls and peaceful domesticity cannot protect us from the arrival of the unknown, can they? And now that same car, to the accompaniment of a waltz, is rushing up a wall in defiance of the laws of gravity. … Higher and higher. Excelsior! The motor car zooms up into the sky, Meteors fly past, but around them is nothing but empty space.[33]

From this we can deduce that Paul's films were reaching a worldwide audience in 1906, and also that *The ? Motorist* may have been accompanied by a waltz at the screening Bely attended in Moscow. But more importantly, that this vein of "realist fantasy" found a contemporary cultural resonance. As an early self-confessed film enthusiast, Bely found in the anarchic violence and fantasy of early films a sense of the fragility of modern life and the impending doom that would color his great novel *Petersburg* (1916). His response to *The ? Motorist* may be exceptional, but it alerts us to the likelihood that there was a wide range of equally personal responses, quite apart from explicit early fiction about film, such as George Sims's "Our Detective Story" and, in very different registers, Kipling's "Mrs Bathhurst" and Apollonaire's "Un beau film."[34]

Melodrama

Very few of Paul's later and longer films have survived, making discussion of their more complex melodramatic narratives difficult. One of these, *The Lover and the Madman* (1905) is known only from three striking images in addition to the catalogue account.[35] A girl is enjoying a country tryst with her lover, until he is called away by a messenger, leaving her at the mercy of an "escaped lunatic . . . who grotesquely makes love to her" before threatening her with a knife. The catalogue image of this moment was reproduced as long ago as 1948, in the first history of early British cinema, and it remains an exceptional image from this period of filmmaking, with the figures in a close two-shot. Another of the images shows what happens next — when the lover returns and the lunatic is killed in their struggle — seen in a much wider shot. From these, it appears that Paul's team was experimenting with an early version of interpolated close shots.

The central part of the film consisted of "an exciting hunt for the murderer . . . depicted in its different phases," which suggests further use of the chase structure, followed by his arrest and imprisonment, before appearing in what may have been one of the first realistic courtroom scenes in British filmmaking. The setting is essentially the same as used farcically in *The ? Motorist*, with a less facetious coat of arms behind the judge. The girl's and messenger's evidence secures the young man's release, and all ends happily with their wedding, apparently involving many cars. Running for nine minutes (576 feet), *The Lover and the Madman* would remain Paul's longest fiction film, and its three-act structure points to an early grasp of dramatic as well as filmic form.

The only complete example of Paul's later fictional output, discovered in Sweden in 2016, uses the same "lunatic at large" premise.[36] *The Fatal Hand* (1907) starts dramatically with a full-screen image of a poster offering a reward for help in catching a four-fingered "homicidal lunatic" on the run from Broadhurst asylum. That name would probably have evoked "Broadmoor" for British viewers, well known as an institution for impris-

62. *The Fatal Hand* (1907). The latest of Paul's dramatic films to be discovered, *The Fatal Hand* opens with an announcement that a homicidal lunatic has escaped from an asylum. Showing the subsequent three murders in a variety of everyday settings follows the pattern of the influential *A Daring Daylight Burglary*, made in 1903 by former Paul employee Frank Mottershaw. Photograph courtesy of the Swedish Film Institute Archive.

oning the criminally insane. But there was also a closer source of inspiration. The massive Colney Hatch asylum on Friern Barnet Road (where *The ? Motorist* opens) was a familiar landmark near Paul's studio in New Southgate, and a fire there in 1903 led to many deaths and a mass escape of inmates. In Paul's film, after an initial murder, the escapee catches a local train and later flees past a building that may be another North London landmark, Alexandra Palace, before being cornered on the scaffolding surrounding a nearby house. Given such a lurid and violent story, the low-key everyday setting follows in the style of Mottershaw's *Daring Daylight Burglary*, which J. H. Martin had evidently adopted in his work for Paul.[37]

It would obviously be a mistake to read too much into such "lunatic" plots, which might be linked more plausibly with the frequent deployment of "tramps" and "gypsies" in early film.[38] These latter figures of no fixed abode operate on the margins of respectable society and are prone to opportunistic and threatening activities. Lunatics or madmen are equally convenient intruders, who disrupt and threaten, thus generating fictional friction. Ever since Edgar Allan Poe's tales had created a Victorian taste for crime and the "gothic" in short fiction, there was no shortage of lurid inspiration for early filmmakers. Cinema would acquire its first detective heroes, Nick Carter and Sherlock Holmes, in the year after Paul's *The Fatal Hand*, once again showing him to be in tune with contemporary taste. Indeed, one of the genre's literary pioneers, G. K. Chesterton, had already celebrated detective stories as the form of modern romance that best expresses "some sense of the poetry of modern life."[39]

Final Narrative Directions

If Paul anticipated the "true crime" genre that would be widely adopted by filmmakers around 1910, there are also signs that he was planning other potential directions for dramatic work—one of which remains particularly puzzling. A trade paper notice from 1908 announced that he would be producing a series of "biblical scenes," beginning with a version of *The Prodigal Son*. This did indeed appear in his catalogue and is illustrated (though lost), but there is no sign of any other biblical subjects being attempted. However, a cryptic reference by the chairman of the British Kinematograph Society meeting that Paul addressed in 1936 refers to his "reformation" period, when his films were shown to the bishop of London. There was at this time a long-running debate over the "Sunday opening" of places of entertainment, which Paul may have been trying to influence on behalf of the trade. But otherwise, this remains an obscure reference to some episode in Paul's career.

The other new direction pointed toward outright spectacle. A fragment of one of Paul's last productions, *The Burning Home*, has recently been identified, and it confirms the impression that this was a major effort on the part of the company, which involved setting fire to a substantial building as the climax of a father's race to save his family. In the fragment, we see firemen rescuing a child from a blazing house on a long ladder, to the relief of his distraught mother. A review welcomed this as a "return to form" by Paul, "after a lapse which has been remarked by the whole of the trade," suggesting that Paul had been holding back from prevailing trends in fiction.[40] It also recognizes the ambitious scale of the production:

> One cannot burn a row of cottages, use 100,000 gallons of water and employ firemen and policemen without paying the piper; in our opinion, the expense was merited by the result, and we recommend buyers to call and see *The Burning Home*.

The film apparently succeeded in gaining American distribution, presumably on the basis of its spectacular quality, amplifying one of film's earliest

63. *The Burning Home* (1909), one of Paul's last films. The spectacular climax of this fire rescue story was recently discovered as a fragment, suggesting that this may have been a "sample," issued to promote sales of what was an expensive production. Photograph courtesy of the Huntley Film Archives.

64. Although the *Bioscope* trade advertisement struck a confident tone in May 1909, Paul would soon conclude that production on this scale was "too speculative."

dramatic tropes, the "fire rescue."[41] But evidently this was not enough to encourage further investment on a similar level. Faced with the double challenge of Pathé's now highly industrialized production, supported by continuing price reductions, and the prospect of Edison's cartel closing the American market to British producers, more than this success would have been needed to justify carrying on at a similar level. A decade after the confidence of Paul's 1898 manifesto, cinema had come to demand more than artisanal ingenuity, as almost all of the early British pioneers were discovering.

CHAPTER 9

"Daddy Paul"

THE CULTURAL ECONOMY
OF CINEMA IN BRITAIN

A cartoon published in a trade journal in 1905 offered an unexpected insight into the small world of the early film business in Britain.[1] Eight men are grouped round a table, on which a tiny figure announces "5d de foot anywhere you like." Beside him is a cockerel, the trademark of Pathé Frères. The figure is clearly a caricature of Charles Pathé, who had just shocked the film world by dropping the price of his output from the industry norm of six pence per foot to five, making his films cheaper than anyone else's. One of the men around the table asks, "What's your proposition Charlie?" This is Robert Paul, recognizable by his dark beard, addressing a man who, cigar in mouth, is just as clearly the American Charles Urban. Also portrayed are Cecil Hepworth, producer that year of the successful film *Rescued by Rover*; A. C. Bromhead, head of Gaumont's influential British branch; and the cartoon's creator, Theodore Brown, who would soon launch a rival trade journal.[2] Paul and Urban are shown as the leading figures in this group portrait of the infant business, and many would indeed have looked to their lead in responding to Pathé's bid to corner the market.

When Paul entered "animated photography" in 1894, he would have seen it as a branch of the new electrical business, since the Kinetoscope he was invited to replicate was battery-powered. A year later, he was developing a non-electrical camera to produce new subjects, after Edison had refused to supply him. By the next year, in mid-1896, he was operating a string of film shows in London music halls—making him a pioneer exhibitor—as well as producing his own films.[3] He was also one of just three suppliers of projectors in Europe, as another pioneer recalled, one of whom (Lumière) would not sell on the open market.[4] Yet Paul's identity at this stage, in the press coverage of his filming *The Prince's Derby* in June, was still that of "a clever electrical engineer."[5]

As we have seen, Paul was quick to appreciate the potential historic

65. Cartoon by Theodore Brown in the *Optical Lantern and Cinematograph Journal,*
September 1905, p. 236. Pathé, identified here by the company's trademark cockerel,
had recently reduced its price to 5 pence per foot of film, causing collective alarm
among English producers, who charged 6 pence. Paul, as the senior English figure,
asks the dynamic American distributor Charles Urban, "What's your proposition
Charlie?" Pathé's price reduction would inaugurate a period of anxiety among the
pioneer English producers that culminated in confrontation between producers and
exhibitors over the Paris agreements of 1909. Brown, meanwhile, would found a new
trade journal in 1907, the *Kinematograph and Lantern Weekly.*

and commercial value of film, at a time when few considered it of more
than ephemeral interest. His offer of films for preservation in the British
Museum was made in July 1896, less than four months after he embarked
on music-hall exhibition and two years before Boleslas Matuszewski published his assertion that film offered "a new source of history."[6] And in
March 1897 Paul became the first filmmaker in Britain to claim copyright
on a film, initiating an interest in this complex new issue to which he
would return ten years later.[7] Although the attempt to float his film business as a limited company in that same month proved premature, the
language of the prospectus was prophetic, referring to "a new industry."[8]

Having created what amounted to a studio in New Southgate in 1898,
Paul opened "showroom premises" in High Holborn and appointed Jack
Smith sales manager for the film company. At the same time, Smith was
also responsible for "the taking of topical films," supported by a team of

assistants who, as we have seen, could produce show prints to a tight timetable.[9]

As his business expanded, embracing equipment and films for purchase, Paul showed remarkable foresight for the industry. He had discovered the need to hire projectionists early in his career as an exhibitor, and with fire an ever-present danger, he trained "limelight men" from the music halls.[10] Ten years later, Paul would take the lead in creating a national training scheme for projectionists, with qualifications issued by Northampton Polytechnic Institute, a North London college that had emerged in 1894 from a charitable foundation originally intended to promote "the industrial skill, general knowledge, health and well-being of young men and women belonging to the poorer classes." No doubt Paul's own educational experience at Finsbury Technical College, and his continuing involvement with technology through the instrument business, had given him a keen awareness of the need for training in what was a new and expanding trade. By August 1907 the *Kine Weekly* judged that "the right key has been struck," in view of the "great number of applications and inquiries" received about "the examination and certification of operators and exhibitors."[11] Two levels of certification, ordinary and higher, had been arranged jointly by the Northampton Institute and a new industry body, which Paul had also been instrumental in launching.

The immediate response to Pathé's 1905 price reduction was that the group of British producers pictured by Brown resolved to form a Kinematograph Manufacturers Association.[12] An early film trade handbook refers to Paul's having "worked so hard in the initial stages of the business," and now "throwing energy into an association to preserve the rights of manufacturers," which suggests he played a leading role in creating the KMA, although, perhaps typically, he does not appear to have served as its chairman.[13] The following year, the association held a public meeting and dinner in October, at which Paul took the chair and reviewed what had been achieved:

> It seems difficult to realise that a little more than a year ago we did not realise the meaning of the word "association" — of course we knew its dictionary meaning, but had yet to realise what a trade association could do for us.
>
> Starting with a meeting in this building, at which the whole of the trade attended, to consider how a threatened war of price-cutting might be obviated, it at first seemed that unanimity was impossible, but the foundation of success was laid when the minority agreed to loyally abide with the decision of the majority, with the result that competition is now chiefly

in respect of quality, rather than in reducing prices to an unremunerative figure. Having arrived at this point . . . it became possible to tackle other problems of vital importance to the industry, such as the correction of various trade abuses, the prevention of piracy of film subjects, the standardisation of film dimensions, and the raising of the standard of quality of film subjects, and of public exhibitions.[14]

Paul ended his speech with a rallying call, looking forward to English films becoming "universally famous for smartness, cleanliness, clearness, steadiness, and high quality."

Establishing Film Copyright

Apart from the traditional exchange of toasts and air of self-congratulation, the main content of this event was the reading of a paper commissioned from a barrister, William Jago, on the subject of film copyright. There is good reason to suppose that Paul had taken the initiative to commission Jago's presentation, given his early interest in this topic. After registering copyright on *Sea Cave Near Lisbon* in 1897, Paul was registering copyright on another film in 1902 when he recorded a dispute with the cameraman. The subject was the Prince and Princess of Wales, filmed by "G. Francis," and Paul's note stated, "I am quite unable to give his full name, legal proceedings being pending between us [however] the assignment to us has been signed and such assignment has been legally approved of."

We might wonder how much of a problem protecting the copyright of early films was. In 1985 André Gaudreault described film piracy as "extremely common" between 1900 and 1906 among all the major production companies in England, France, and the United States. "All producers at the time enthusiastically pirated (by duping a print) the films of competitors who had not taken the precaution of copyrighting them at the Library of Congress."[15] But Richard Brown has challenged Gaudreault's "sweeping statement," suggesting that "whilst film 'duping' probably existed in England before 1912, it was certainly not the widespread activity that it was in the United States."[16]

This was almost certainly because the British film trade was comparatively collegial and highly concentrated in London's West End, with similarly concentrated regional distribution centers. Certainly there had been extensive unauthorized copying of films in the earliest period: Edison copied Acres and Paul's first films, notably *Rough Sea at Dover*, which was the hit of his 1896 Broadway debut screenings.[17] And Paul may have copied Edison's films, or the purchasers of his replica Kinetoscopes could have done so. But by the early 1900s, systematic piracy would almost certainly have been detected—as it was by Charles Urban, who threatened prosecution of London exhibitor Walter Gibbons, and a 1903 case in the United States brought by Edison against another producer, Siegmund Lubin. It was in this context of producers concerned to protect their investment that Jago presented his paper, "The Law of Copyright in Kinematograph Pictures."[18]

Jago started by distinguishing two senses of copyright: "the right which a man has to control, and if he wishes, to prevent the publication of his literary publications," and "the exclusive right of multiplying the copies of such production after publication." Once published, a work was protected by the Copyright Act of 1842, although as Jago admitted, "There is nothing legally literary about a kinematograph picture." There would be redress under common law if someone was shown to be "passing off" their work as that of another. But what protection might there be for the subject or performance embodied in a film?

Jago cited the historical case of *Turner v. Robinson*, dating from 1860.[19] Robert Turner was a print seller in Newcastle, who had bought the engraving rights to Henry Wallis's hugely popular painting *The Death of Chatterton* (1856). James Robinson, of the Polytechnic Museum and Photographic Galleries in Dublin, was one of the first generation of stereographic publishers to cash in on the new fashion for stereoscopic views. He had published a stereograph of "The Death of Chatterton" in 1859, which Turner claimed was a "piratical imitation" of the Wallis painting.[20] We learn from a contemporary account that the case had come about when Turner shipped the painting to Dublin in order to attract subscriptions for his engraving, whereupon Robinson had announced that he would soon offer "the beautiful stereoscopic figure of the last moments of Chatterton."[21]

When Turner applied for an injunction, Robinson's defense was that "his stereograph was not copied from Mr. Wallis's picture, but was an independent study from the biography of Chatterton," and that he had "from recollection built up the subject in his studio and made a stereoscopic photograph of it." At a second hearing, he further argued that "it is impossible to take pictures for stereoscopic slides from a plain surface such as a picture."[22] Despite this, Turner succeeded in getting an injunction against Robinson. While "the stereographic slides are not photographs taken directly from the picture, in the ordinary sense of copying," the court ruled, "they are photographic pictures of a model, itself copied from and accurately imitating in its design and outline the petitioner's painting."[23]

Jago pursued at length two issues that continued to perplex those seeking to apply statutes codified before the development of moving pictures: first, whether a sequence of photographs could be regarded as a picture and, second, whether the representation of a performance can be treated *as* a performance. On the first, he argued that although the negative of a film consists of a series of pictures, "their sole use is for the purpose of producing when exhibited one picture only." As to the second, he suggested that "no one who has heard the roars of laughter caused by a comic moving picture can doubt that a Kinematograph exhibition may certainly cause the same emotions as those produced by a representation by actors of the same scene or event."

Although he did not mention it, Jago may well have been aware of the 1903 US Circuit Court ruling in *Edison v. Lubin* in 1903. This has often been characterized as one pirate taking another to court, although in truth American copyright law at this time was radically unclear about the status of film. Lubin's defense against Edison's charge of piracy was that Edison would have had to register each frame of his films individually to secure protection under the US Copyright Act of 1870. A lower court agreed with Lubin, but its decision was reversed by the appeals court, which held that "while the advance in the art of photography has resulted in a different type of photograph, yet it is none the less a photograph."[24]

Citing judgments in which "scenic entertainments" and "producing emotions" had been regarded as criteria for something being considered a "dramatic representation," Jago concluded that in Britain "a kinematograph representation can claim the protection afforded by the Dramatic Copyright Act of 1833, while the Fine Arts Act protects the film itself from all trafficking, whether by copying, multiplying or exhibiting." But whatever satisfaction the KMA members might have felt after this learned analysis of their situation was about to be rudely disturbed.

The landmark case *Karno v. Pathé Frères* would turn on the impresario Fred Karno's claim that the Pathé film *At the Music Hall* infringed his copyright on the music-hall sketch "The Mumming Bird," which he had registered in 1906.[25] After the court visited the Oxford music hall to see the sketch performed, and viewed the film, the judge found that the film was indeed "a representation of the plaintiff's sketch" but that this sketch did not qualify as a "dramatic or musical performance." Although the judgment was a blow to Karno and left open the question of who might be liable if such a case *were* proved (the producer or the exhibitor of the film?), it established that no "legitimate" dramatic work could be filmed without the author's permission. A parallel case in the United States, *Harper Brothers v. Kalem Company and Kleine Optical* (1908), which arose from Kalem's unauthorized filming of Lew Wallace's *Ben-Hur*, already a popular stage and arena entertainment, turned on whether Kalem's dialogue-free adaptation of a literary work could be considered a "dramatization." In an eventual Supreme Court ruling delivered by Justice Oliver Wendell Holmes Jr., Kalem's screen version was judged to infringe Harper's copyright, thus establishing the rights of the owners of the source-work of a film.[26]

Later in the same year, an international copyright conference in Berlin revised the original Berne Convention of 1886, taking account of its 1896 Paris revision and including for the first time an article that referred to film. In Britain, a new Copyright Act followed in 1911, which incorporated the Berlin provisions, thus giving film formal copyright status for the first time.[27] America, meanwhile, continued to reject the principles underlying the Berne Convention until the later twentieth century.[28]

Paul was active throughout the decade in commenting on technical matters of common interest to the film business. He contributed articles to *Kine Weekly*, addressing, in 1907, moves to standardize the dimensions of the film strip and its perforations and, in 1908, what he believed was the neglected potential for using film in technical education.[29] That he did so at a time when the British film industry was in violent upheaval probably points to a growing personal detachment from the commercial future of the industry. His role as an elder statesman was still in evidence in January 1909, when he deputized for the KMA chairman, T. C. Hepworth, at another trade dinner. In his speech, Paul spoke of the history of "boom and slump" that he had known over twelve years. He also recalled the excitement of finishing the first film he had made at three o'clock in the morning, with a cheer loud enough to attract a passing policeman—the incident that would later be attached to William Friese-Greene.[30] He hoped that "brains and a variety of thought" would help the trade through the present difficulties, allowing individualists to thrive and avoid being taken over by "capitalists and company promoters." In light of the storm that had broken out over new terms of trade—and Paul's imminent departure from the industry—it is hard not to read this as a final appeal. Meanwhile, his rejection of holding a debate at the dinner was accepted, with a date set in February to discuss the crisis.

A Trade Revolution

Throughout 1908 and 1909, the British trade press carried numerous articles and letters from readers, all invoking a sense of crisis or "revolution." There were in fact a number of developments under way, essentially unrelated but contributing to a sense of unease and growing antagonism between different groups within the industry. One ongoing factor was Pathé's lowering of prices: the company had followed its 1905 cut from six pence per foot to five with a further reduction to four pence in December 1906, and three pence per foot being discussed early in 1909. Other factors were the rise of trade unionism within the exhibition sector (a National Association of Cinematograph Operators was formed in 1909, drawing support from the example of a successful strike by the Variety Artistes Federation in 1907); a continuing debate over allowing the Sunday opening of cinemas; and pressure by the KMA for government legislation on fire safety, which would eventually lead to the Cinematograph Act of 1911 — with unforeseen consequences that triggered the creation of a self-regulating film censorship board.[31]

But in 1908, the main controversy stemmed from the "fundamental market antagonism of makers and exhibitors."[32] There were in fact three sectors at odds with each other: makers, or producers; renters, later known as distributors; and exhibitors. The producers had been the first to create a trade association, but both renters and exhibitors would soon follow their lead. According to Rachael Low:

> The essence of the problem was that renters and exhibitors, having bought films from the manufacturers, kept them in circulation too long . . . thus depressing prices. The result was a continued effort on the part of manufacturers to limit the length of a film's life.[33]

These issues prompted two international industry meetings in Paris, one initiated by the British KMA, held on 9 March 1908, and an "International Congress of Film Makers and Publishers," held in February 1909 and attended by representatives of thirty companies based in seven countries.[34] Ten British production and distribution companies were represented by negotiators from the KMA, among them Robert Paul. The main terms agreed were that renters and exhibitors would sign a contract that committed them to maintaining a minimum price per foot, returning all films to makers after four months, and not obtaining films from nonconvention sources.

Opposition to these terms broke out immediately and vociferously among English renters and exhibitors, voiced mainly through the *Bioscope*, which denounced "would-be monopolists" and "grandmotherly dictation."[35] Moves to unify this opposition led to the formation of the Cinematograph Trade Protection Society, which attracted mainly smaller companies but also some of the larger renters, such as Jury and Ruffell.

66. Group photograph taken at the Paris Convention of Filmmakers and Publishers, February 1909, organized to promote unity among European producers, faced with Pathé's price-cutting and the threat of Edison creating a cartel that would exclude them. Georges Méliès chaired the meeting (front row, fourth from left), which Charles Pathé attended (front row, second from left). Paul and Cecil Hepworth appear at the extreme right of the fourth row.

These renter-distributors carried weight within the industry in part because British exhibitors were increasingly turning to foreign imports, finding that British producers were not offering films on a scale to equal those from France, Italy, and increasingly America. This was the era that saw "headline" films become "exclusives," and lengths increase beyond 1,200 feet, with lavish productions from, especially, Italy commanding high prices.[36] *Kine Weekly*, in contrast, tended to support the KMA's position, defending English films as having "merits that are not present in the foreign product," and criticizing English showmen for showing "films of very slight merit with monotonous regularity" rather than "the real 'headliners'" on offer from English producers.[37] Pathé, the largest of all producers at the time, with a network of affiliated companies in many countries, stayed aloof from the decisions of the Paris Convention. And Edison would shortly create a new cartel, the Motion Picture Patents Company, which had the effect of spearheading American exports while closing off the American market to nonmembers. A new organization was formed to represent English producers in America early in 1909, the International Projecting and Producing Company, which Paul joined, but this did little to restore its members' market share.

In Britain, new and mainly short-lived trade bodies emerged during this tense period: the CTPS gave way to the Cinematograph Defence League, which in turn was replaced in 1912 by the Cinematograph Exhibitors' Association. And during 1911, the KMA reached an agreement with a new organization of larger renters, the Incorporated Association of Film Renters, to try to impose control over the market. Meanwhile, more fundamental changes were taking place across the industry, with a shift toward longer films, which would become the "feature film" standard that has survived until today. And with features came a new pattern or renting under "exclusive" terms, which guaranteed exhibitors that they alone would have a new film in their area.

Paul's Departure from the Film Industry

By the time these changes took place, Robert Paul had quietly quit the film business. According to Rachael Low, Paul's "formal retirement from film production" took place in 1910, although no source for this was given, and detailed searching of the trade press has not revealed any announcement. Paul's name simply disappears from the weekly listing of new releases near the end of 1909. The only statement he appears to have made came in 1936, at the end of his BKS paper:

> By 1910 the expense and elaboration necessary for the production of a saleable film had become so great that I found the kinematograph side of the business too speculative to be run as a sideline to instrument making. I then closed it down and destroyed my stock of negatives, numbering many hundreds, so becoming free to devote my whole attention to my original business.[38]

Taken at face value, Paul's assessment of the situation in 1909–1910 can hardly be contested. Most of his fellow pioneers had effectively ceased production, with only Cecil Hepworth, and for a short time Will Barker, scaling up their production to meet the new challenges. Charles Urban's solution was to invest heavily in Kinemacolor, which proved a sufficient novelty to allow him to continue producing nonfiction on a larger scale.[39] Paul's decision to quit production, in face of a rapidly changing market, appears reasonable and indeed shrewd. And unlike his contemporaries among the English pioneer producers, he had another substantial business to manage.

However, several aspects of this abrupt and unheralded move are still perplexing. One is the fact that Paul appeared to be increasing the scope

67. Paul's last appearance in the *Kine Weekly*, 28 October 1909, acknowledged his pioneer status, with no suggestion that he would shortly abandon the film business.

of his production at the very point of quitting. His last major film, *The Burning Home*, had been reviewed encouragingly in *Kine Weekly* in May 1909, while the outlay involved in making it had also been noted.[40] Presumably the commercial results of that and other late films had been less encouraging and persuaded him, as he later said, that this scale of production had become "too speculative," even as the trend toward longer exclusives reduced demand for nonfiction, hitherto a staple for him. Or maybe the vitriolic response by exhibitors and smaller renters to the terms of the Paris Congress left him discouraged.

We will never know which of these was the deciding factor. But neither is it easy to understand how a figure as prominent as Paul within the British film industry could abandon his entire business without any comment appearing in the trade press. Had he taken steps to somehow suppress any discussion? Paul appears to have been intensely private and, as he later noted, "averse to personal publicity," at least after his *annus mirabilis* of 1896. But wouldn't closing down all his film activities so abruptly have given rise to employment issues or other visible repercussions?

Perhaps the single most surprising, and inexplicable, aspect of Paul's exit from film is the destruction of his "stock of negatives." Paul had been in the forefront of trying to guarantee film some permanency, from the early offer of films to the British Museum through to his donations of equipment to the Science Museum and the Royal Institution. So why would he destroy negatives? Such an extreme action can only suggest that more than an economic judgment underlay his decision to abandon film production. Since we now know that the death of his son, seemingly due to a film-related accident, was ten years earlier, we are left to contemplate some other emotional reason for his decision to do something so seemingly uncharacteristic as destroying the legacy of his fifteen years work in film.[41]

CHAPTER 10

"My Original Business"

PAUL'S TECHNICAL AND SCIENTIFIC WORK

When Paul, in 1936, described his film career as a "sideline" to his "original business" of making electrical instruments, he echoed two other major figures in film history, Thomas Edison and Louis Lumière. Neither of these appear to have considered film production their greatest achievement, and neither maintained an extended connection with the medium they had largely launched.[1] As we have seen, Paul was different in that he remained practically involved in his film business for at least fifteen years, while also managing a successful instrument business until at least 1919. It was then that he merged the R. W. Paul Instrument Company with the older Cambridge Scientific Instrument Company, which became the Cambridge and Paul Instrument Company until, in 1924, it was floated publicly and renamed the Cambridge Instrument Company. Paul is described as having withdrawn from instrument design, "concentrating on financial matters," but he also remained a resourceful hands-on inventor, as several ex-employees testified, and an active campaigner for science education.[2] Indeed, when we begin to connect the diverse strands of Paul's activity from 1910 until the late 1930s, it becomes clear that the same qualities which had made him a successful film pioneer continued to motivate him.

It was as an instrument maker that he had taken premises in Hatton Garden in 1891, before the fateful appearance of Georgiades and Trajidis in 1894. The business was apparently prospering and had already required him to take on additional premises in nearby Saffron Hill.[3] In 1892 he published a catalogue of electrical testing instruments and patented the galvanometer that he would advertise the following year—and that would become his most successful product in a wide variety of versions.[4] The instrument historian C. N. Brown has noted that the patent did not mention either of Paul's teachers, William Ayrton or Thomas Mather, but he observes that "presumably there was some agreement between the three men regarding the manufacture and sale of the galvanometer." Film historians might infer from this that Paul already had experience in agreeing issues of intellectual property uncontentiously before Birt Acres's notorious falling-out with him.[5]

In view of Paul's eventual rejection of the film business, it may also be worth noting that much of what he was selling at this time was "general educational science apparatus," aimed at the new and growing market for electrical education from which he had himself benefited at Finsbury Technical College. In his 1932 correspondence with the German pioneer Oskar Messter, he would say of the original Theatrograph projector, "My desire was to produce an attachment for existing lanterns, chiefly to be used in Schools and Colleges, and to be sold for £5."[6] This could perhaps be a later rationalization, or perhaps an insight into Paul's initial motivation in 1895. But it is clear that, even as both his instrument business and the film trade were flourishing around the turn of the century, he remained personally committed to continuing research and development. We have seen how the structure and variety of his film productions flourished at this time. Meanwhile, Paul was also applying his engineering and commercial skills to the apparatus of film, with a string of improvements to projectors and other equipment. As early as 1900, he was advertising a "New Century Animatograph," with "lanterns, lenses, jets, limes [limelight], arc-lamps, resistances, ammeters, repairing outfits, and all accessories for animated photography," kept in stock at his Animatograph Depot in High Holborn.[7] In the 1901–1902 catalogue, the largest he ever produced, a section headed "Experimental Department" shows a well-equipped work bench, with an invitation:

> My endeavour being constantly to improve the production of animated photographic apparatus, I have devoted a special workshop to this, and shall be pleased to receive suggestions from any of my clients, whether such are patented or not, and to take up suitable inventions on favourable terms.

The 1902 catalogue had eight pages detailing Paul's latest version of the Animatograph projector, with diagrams showing how its unusually small shutter reduced flicker and provided "extreme brilliancy of the picture on the screen." A further twelve pages laid out the full range of accessories Paul offered, ranging from arc lamps to a new Animatograph camera, with a "revolving tripod head," to developing, printing, and perforating equipment.[8]

68. Photograph from Paul's 1901 catalogue, illustrating the "experimental work-shop" of Newton Laboratories at his "new film works" in New Southgate (now Mus-well Hill). Here Paul appears to have followed the same principle that informed his instrument-making business, based on a constant process of product improvement and innovation.

Another page repeated an offer made in Paul's early catalogues to take "local and special subjects" according to customers' requirements, which suggests that this continued to form a part of his business, although there is no way of knowing who may have commissioned films. Nor is much known about Paul's science-related work in this period except from his 1908 article "The Kinematograph in Technical Education," in which he describes the production of two series of demonstration films for leading scientists, no doubt customers of his instrument business.[9] As a historian of Cambridge Scientific Instruments noted, like that company's founder, Horace Darwin, "Paul preferred to devote time to the more interest-ing work of making specialised instruments for his friends as an aid to their researches," adding, "fortunately, he could afford to subsidise this work" — no doubt at least initially from his film earnings.[10]

The scientific demonstration films may be rare examples of these two strands of Paul's work and interests coming together. One series was com-missioned and possibly supervised by R. W. Wood, a distinguished Ameri-can physicist, then at the University of Wisconsin.[11] Wood published an important paper in 1900 on visualizing sound waves using the cinemato-graph, which may have led to his collaboration with Paul.[12] The film is de-

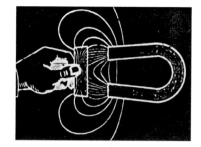

69. "Specimen diagrams" from two series of films made by Paul "under the direction of Professor R. W. Wood of Wisconsin and Professor Silvanus P. Thompson of London." The first, filmed at 120 frames per second, illustrates a sound wave reflected in a spherical mirror. The second, "magnetic lines of force, first imagined by Faraday as a means of 'picturing and solving various magnetic phenomena.'" A hand is shown removing the keeper from a permanent magnet, "and as the hand pulls it off the forms of the lines change." "Thus by a film formed by photographing several hundreds of such diagrams it is possible to give the student a moving picture as imagined by Faraday." Robert W. Paul, "The Kinematograph in Technical Education," *Kinematograph and Lantern Weekly*, 3 December 1908, 273.

scribed as illustrating "the changing forms of a sound wave, which originating at the focus of a spherical mirror, strikes it and is reflected." Paul explains how he created "actual photographs of sound waves by means of an electric spark," due to the fact that "in travelling through the atmosphere the wave causes alternate rarefaction and condensation of the air, so that when a beam of light strikes such layers it undergoes refraction," producing a shadow that can be photographed, albeit only by a camera running at 120 frames per second.

For what must have been a spectacular version of this, a canister of dynamite was exploded in a chalk valley, which provided "the required light background."

> The camera having been started, recorded a dark ellipse, gradually growing in eight successive pictures in which it is plainly shown [but] since its radius is growing at 330 metres per second it is soon too faint and distant to be visible.

A cryptic reference to "the combination of phonograph and kinematograph" suggests that an audio as well as a visual recording may have been made of this experiment, although this is not fully explained, except to make a point about how spectators at different distances from the screen would have different experiences. Having built a high-speed camera, Paul clearly wanted to exploit its potential for other purposes, and he de-

SILVANUS P. THOMPSON LECTURING TO A JUVENILE AUDIENCE AT THE ROYAL
INSTITUTION, CHRISTMAS 1910.

70. Silvanus Thompson lecturing at the Royal Institution in 1910. A major figure in the development of science education in Britain, Thompson (1851–1916) was principal of the Finsbury Technical College from its foundation in 1878 to 1908. A renowned lecturer, he gave the Royal Institution Christmas Lectures twice, in 1896 and 1910, probably using Paul's animated film of "magnetic phenomena" for the second series.

scribed an experiment carried out under the direction of Vernon Boys, a physicist known for his work on calorimetry but also for his popular exposition of the properties of soap films.[13]

Another series of films was made for Paul's mentor at Finsbury Technical College, Silvanus Thompson. These were to show the "so-called magnetic lines of force, first imagined by Faraday as a means of picturing and solving various magnetic phenomena." Since these are theoretical, "they cannot of course be photographed," but Paul had produced an animated film "of several hundred of diagrams ... to give the student a moving picture as imagined by Faraday." Again, there is a suggestion that this illustration went much further, showing "the much more complicated magnetic circuits of alternating current machinery" in a series of six films. Paul notes that these short films could looped as "endless bands," allowing them to be viewed repeatedly, and no doubt with a running commentary by the scientist. Unfortunately, none of them are known to have survived, although as Paul observed, "their educational effect is considerably greater than that of a film of Irishmen dancing, which has been advertised as part of a geographical series." Here we may wonder if Paul was referring to a film recently released by his own company, *Whaling Afloat and Ashore*, which includes a sequence of the Irish whalers dancing together.

Could this be an early indication of his impatience with the market's demand for entertainment? The article ends with a rather clear expression of this sentiment:

> The commercial field for subjects of this nature must naturally remain small, but there are still many directions in the fields of engineering and constructive work in which Kinematography still needs development, though up to the present its use for amusement purposes has been paramount.

Just over a year later, Paul would abandon the world of "amusement," still apparently frustrated by how little the potential for using film in science and technology education was being exploited—despite Urban's apparent success in this field and the repeated, though unanswered, calls of leading educationalists for greater use of film in teaching.[14]

The Unipivot Breakthrough

Progress of Paul's instrument business is almost as difficult to track in detail as his film work. We know that his instruments were awarded gold medals at two international exhibitions, the Saint Louis Exposition of 1904 and the Brussels International Exhibition in 1910. But his greatest and most enduring success came in 1903, with the launch of his first unipivot galvanometer, which was allied to a new mode of manufacture. As C. N. Brown observed

> Before 1903, Paul was a traditional instrument maker using craftsmen to make individual instruments in the time-honoured manner. After 1903, he was manufacturing instruments, almost mass-producing them, in a much more modern way, assembling dozens of variants and accessories from a range of standardised parts.[15]

To understand what Paul had achieved involves recalling the rapid growth of electrical industries in the late nineteenth century. Telegraphy had been one of the first applications, along with the technique of electroplating. Both required the ability to measure electrical current, as did lighting and other applications that followed with the technology of power generation. A galvanometer is essentially an instrument for detecting and measuring current, based on the interaction of an electrified wire coil and a permanent magnet, and by the 1880s, there was growing demand for a range of these to meet different requirements—accuracy

71. Unipivot galvanometer, 1917. Patented in 1903, Paul's design used a single pivot for the moving coil, which minimized friction and maximized sensitivity in measurement of currents. Robust and convenient to service, it won a gold medal at the Louisiana Exposition in 1904 and remained in production through the Cambridge Instrument Company for over fifty years. The "unipivot" principle was used in other Paul measuring instruments, becoming his most distinctive contribution to instrument design. Photo: Ian Christie.

and portability being the most important criteria. Paul's teachers at Finsbury College, Ayrton and Mather, had created an optimum version of the moving-coil pattern, while William Thompson (Lord Kelvin) had developed a more sensitive moving-magnet design. Both, however, needed careful handling and calibration, and were subject to distortion from external sources. It was the American Edward Weston, already a veteran of several electrical businesses, including plating, power generation, and lighting, who succeeded in creating a moving-coil meter in 1888 that became the basis for voltmeters, ammeters, and the new generation of instruments increasingly in demand.[16]

In Weston's model, the coil was mounted between pivots, with its movements controlled by springs. This worked well for measuring larger currents and voltages, but less so for smaller values. Paul's "improvement," as his patent was titled, involved mounting the coil on a single pivot and controlling it with one spring. This increased sensitivity, and having the pivot at the center of gravity gave it greater robustness in the event of tilting. As indicators of the importance Paul attached to the principle, "unipivot" became his telegraphic address, and Newton, the house he built on Sydney Road in Muswell Hill in 1914, was fitted with custom-made window catches that ingeniously incorporate the unipivot circular bearing design. Brown speculates that Paul may have introduced the unipivot system because he realized Weston's patents were expiring.[17] Whatever the motivation or inspiration, the design was enough to bring Paul the St. Louis gold medal and to launch a range of unipivot-based instrument variations that would continue in production for the subsequent fifty years. In 1913 he introduced an adapter that made possible the measurement of alternating current (AC), as well as the direct current (DC) that was still widely in use. After the "current wars" in 1880s America, between proponents of AC and DC—the latter championed by Edison—new generating systems increasingly turned to AC. That Paul opened a branch office in New York in 1911 may be linked to his development of the AC adapter.

War Work

The outbreak of the First World War in August 1914 would dramatically expand the range of Paul's manufacturing and development, and eventually bring an end to his commercial independence. Although detailed information is lacking, it seems certain that he was involved in producing shortwave wireless signaling equipment for use in the trenches, on behalf of the War Office's Experimental Signals Establishment.[18] Another war-related product was a device to measure the altitude of aircraft, the Paterson-Walsh height finder, as aerial operations became increasingly important.[19] And on the marine front, Paul was active in proposed systems for detecting submarines and mines.

Thanks to others involved in these activities, there are several revealing anecdotes about Paul's unorthodox behavior on official occasions. And there is also the remarkable text of a formal speech he gave to employees at the Muswell Hill factory on the first Monday after Britain had declared war, in which, with considerable foresight, he anticipated an extended conflict:

We resume work under conditions so changed that it is scarcely possible to focus them. We, who have not lived the daily, military life of our powerful continental enemies and allies, find it impossible to gain an idea of the forces now at work, which must, in any event rupture many of our commercial connections and for the remainder of our lives change our entire outlook. Practically all French and German trade is cut off, and all their able-bodied men are at the war.

We, as a nation, have awoke to the fact that security cannot be bought by the payment of professional soldiers, and that, if we are to hold our place in the world, all slackness, love of ease and selfishness must be thrown aside. The responsibility for the war is not upon our consciences, but we must reap its result from now onwards.[20]

Paul then outlined "our respective duties under these altered conditions":

For my part, I conceive it my duty to carry on business as long as work, materials, and money for wages can be in any way procured.... A definite amount of stock should be now in each store, and I leave in a few moments to endeavour to secure more. A certain amount of work has been assured by Government contracts just placed, and more may follow from the same source if satisfaction is given.

The work specified was "signalling apparatus for both the Army and Navy," and staff were reminded that this "will be done under the Official Secrets Act, and to divulge information respecting it is treason." Next, he set out a series of conditions for employment during the war, covering notice to be given if staff had to be reduced — with none for any who are "unpunctual, negligent or disobedient" — and the promise of jobs being kept open for those who enlist, while urging others to "serve by enrolling in the special constabulary." "War," he observed, "is not only a trial of moral character and of courage, but an engineering and scientific contest in which we can play our part."

Somewhat surprisingly, considering that Germany was the enemy in the newly declared war, Paul then reached into the history of scientific instrument making for a parallel:

One recalls that during the Franco-Prussian war the founder of the great optical firm of Carl Zeiss turned his hand to repairing rifles with one assistant, and that they lived on bread, flavoured with an onion or an occasional herring.

The speech ended with an appeal for "working cordially together" in the interests of securing continuous employment, with as little supervision as possible "in order that office staff may be free" to concentrate on "providing material and money." Then a final exhortation:

> I would ask you to regard my remarks, not as those of an alarmist, but as an indication of my desire to respect our future welfare.

Paul's wartime work undoubtedly involved cooperating with technologists and scientists who were already part of his professional network. One of these was Clifford Copland Paterson, ten years younger than Paul but born in the same area of London and also a graduate in electrical engineering of Finsbury Technical College.[21] Paterson had helped devise an airplane height finder, which had been manufactured by Paul and was to be tested competitively alongside other designs. Paterson's account of the test, held at the National Physical Laboratory, is probably the most vivid we have of Paul's manner and attitude to serious work:

> All the height-finders, except Paul's, were erected on the field and got into working order by their sponsors in the morning; but Paul's failed to turn up, and when two o'clock came and the "Brass Hats" began to assemble, it seemed clear that something must have gone wrong.
>
> At a quarter to three a cloud of dust appeared and with it Paul's disreputable Gladiator car, driven by himself and with half-a-ton of height-finder on the back—depressing the back springs to the axle. The impact of the car on the boundary to the demonstration field caused it to bounce into the air and continue to bounce and creak in a hair-raising way, as it risked its heavy load to the demonstration site. Here, with no particular care, the height-finder was dragged down and by three o'clock it was set up and ready for trial.
>
> There was more than a suggestion of remonstrance at taking such risks on a critical occasion, but Paul's only response was that if the thing had to be handled in practice by Army personnel in the field it was just as well to test it for reliability as well as for accuracy. The gear was found to be in good adjustment, and this combined with Paul's demonstration and supreme confidence in his own design, resulted in the adoption of this height-finder for use at Field A.A. stations in the last war.[22]

M. J. G. Cattermole's comment, that "even in the gravity of war the showman in Paul could not be completely submerged," obviously refers to his recently abandoned film career. But knowing a film such as *The ? Motorist*, and even earlier, the anecdote about him filming the Derby in 1896, we might see his impatience and "supreme confidence" as more fundamental character traits.

Another anecdote from a scientific colleague underlines Paul's pragmatic approach to experimentation. Albert Wood, who became a world expert on underwater sound, was working at the Admiralty Experimental Station near Harwich in the later stages of the war, and he recalled Paul being "in and out of the laboratories at Parkeston Quay" testing a system he was developing for a non-contact mine:

> There was the occasion when he borrowed a slop-pail from the Alexandra Hotel and carried it undisguised to Parkeston Quay! He used it to contain his magnet system, to be tested, submerged, with a ship running past. Unfortunately, on the first test, the watertight joint under the lid of the slop-pail failed, and his lovely pivoted magnet system was put out of action.[23]

Wood's memoir records that the scientific head of this station was W. H. Bragg, who had been awarded the Nobel Prize in physics, together with his son Lawrence Bragg, in 1915. Paul would probably have first met William Bragg during his Parkeston Quay period, and they must have stayed in contact. Twenty years later, they would collaborate on producing a portable alternative to the "iron lung," known as the Bragg-Paul Pulsator.[24]

Merger

Although Paul continued to operate his factory throughout most of the war, it was requisitioned by the War Office Signals Department early in 1918 to manufacture trench telephone equipment.[25] He then moved into a former electrical showroom in Muswell Hill, taking those of his staff who remained after conscription to continue experimental work. Another memoir provides a valuable firsthand account of this transitional period. G. Watchorn started work as an apprentice early in 1918 but had little training until the company moved back into Sydney Road the following year:

> War Department work had ceased abruptly, the workers had been sacked, and the place was a shambles. I was told to wire brush the rust off some pipes in the test shop. I saw Paul and said "this is not what I have come

72. William Henry Bragg (1862–1942) was a distinguished British-born physicist who began his career in Australia and in 1915 became the only Nobel laureate to share the award with his son, Lawrence, for their work on X-ray spectrometry applied to the structure of crystals. Paul must have met Bragg while working on submarine detection during World War I, and later collaborated with him on the Bragg-Paul Pulsator.

here for." He agreed, but explained the problems, adding "You must have other work." My next job was to help re-wire the clock system, a master electrical pendulum clock which controlled clocks in all main departments.

Watchorn explains that the factory, dating from 1903, had originally had its own power supply, with a gas engine driving an electrical generator, but that the War Department had introduced changes. The plant was effectively operating on AC current, while maintaining a large battery system "to supply DC current for electrical testing." Paul was visiting auctions at former War Department factories "and buying anything he thought suitable for the factory." He had an office in the main building, "whilst the test shop and unipivot were in 'H. F.,' a building so named because it [had] accommodated the making of Height Finders, an instrument Mr Paul had invented."

This period of recovery was abruptly interrupted by Paul's announcement that he was selling his company to the Cambridge Scientific Instrument Company in November 1919. In its suddenness, the decision recalls his earlier abandonment of the film business, but in this case, the change was a matter of uniting two specialist companies with similar trajectories, both created by exceptional individuals who were initially hands-on instrument makers. However, the founder of the Cambridge company, Horace Darwin, had a considerably more illustrious background than Paul. As the youngest surviving child of the great naturalist Charles Darwin, he enjoyed a privileged upbringing and education. But after graduating from Trinity College Cambridge in 1874, he embarked on an engineering apprenticeship and designed the first of many instruments, a klinostat for recording the rate of growth of small plants.[26] In 1895 he formed the Cambridge Scientific Instrument Company Ltd. with a partner, specializing in anthropometric instruments. Unlike Paul, he received a number of honors, being elected a fellow of the Royal Society in 1903 and knighted in 1918, probably for his work as chairman of the wartime Air Invention Committee, which may well have brought him into contact with Paul.

It seems likely that both Paul and Darwin anticipated a postwar slump in business, which may well have encouraged their merging. And indeed, according to Watchorn, "there was a general trade slump, and no one seemed to want our instruments."[27] The merged enterprise became the Cambridge and Paul Instrument Company, and William Apthorpe, head of the Cambridge company's test room, moved to North London to act as manager. Business gradually improved:

> Apart from a general mixture of high sensitivity galvanometers, Wheatstone Bridges and prototype jobs, there was a demand for standard measure of electrical values . . . made to National Physical Laboratory approved design. . . . Our customers were the developing countries and universities.[28]

Paul had withdrawn from day-to-day involvement, taking a seat on the board of what was now a public company, but he maintained his own workshop in the large house on Addison Road, where he was now living. Watchorn was selected from the factory workforce to work as a technical assistant, and he again supplies a vivid account of Paul's far from relaxed retirement:

> He was working on designs for a petrol gauge, trafficators and a sun visor for fitting to the new Rolls Royce chassis. . . . He had tremendous "drive" when working on a project; he seemed to press on at all hours until it was completed. I came at 8am and left at 8pm. He would be working already when I arrived and would then go off for breakfast. Judging from all the

fag ends lying around, which I swept up, he must have gone on working well into the night.

Watchorn's period of working directly with Paul must have been in the mid-1930s, since he includes in the current projects "improvements in the 'iron lung,'" which occupied Paul throughout much of the decade.[29]

The Breath of Life

The story behind the invention that linked Paul with the distinguished physicist Sir William Bragg — an efficient portable respirator — began with Bragg trying to help an elderly neighbor who was suffering from muscular dystrophy and attacks of respiratory paralysis. The only solution was manual ventilation, which was proving a severe strain on the man's wife and nurses.[30] Bragg's first solution was to connect two football bladders, so that squeezing one would inflate the other, strapped to the patient's chest, and so force exhalation. While the principle seemed sound, the apparatus clearly needed refinement, especially if it was to be used over a long period. Bragg turned to Paul, whom he would have known since their work in wartime research, and also through the Royal Institution, where Bragg held a chair and Paul was an active member. Paul's solution was to create a hydraulic version that could be connected to a water supply if, as was likely at this period, no electricity was available. A rubber bag shaped as a hollow waistcoat was fastened round the patient's chest and connected to pulsating bellows, thus requiring no external assistance. Apparently, the prototype worked successfully every day for three years, apart from one winter when the water supply froze.

Improvements were introduced after Phyllis Margaret Tookey Kerridge, a young physiologist at the London School of Hygiene and Tropical Medicine, was invited to review it in 1934.[31] Kerridge proposed replacing the "waistcoat" with a canvas airbag, and suggested a smaller version for use with newborns. She also fed back information to Paul about improving the measure and control of ventilation. The first and most famous photograph of the apparatus, now formally named the Bragg-Paul Pulsator, showed her laboratory assistant wearing it, and she encouraged Paul to publish an account in the medical press to publicize it, while also notifying the *Lancet*.[32] In a 1934 letter to Bragg, she wrote, "I am sure that it would interest a large number of general practitioners, who would not otherwise hear of it."

The Pulsator did indeed interest many, and proved particularly useful at a time when the Drinker respirator, popularly known as the "iron lung,"

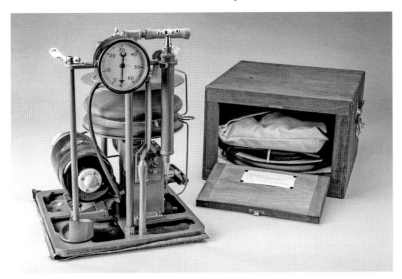

73. The Bragg-Paul Pulsator, developed by Paul at the request of William Bragg in 1935, while the latter was director of the Royal Institution. Deliberately simple in design, the Pulsator was a ventilator for use with patients unable to breathe for themselves, and could be water-powered where electricity was not available. Science Museum.

which had been invented in America in 1927, was both cumbersome and prohibitively expensive. Bragg and Paul's inexpensive apparatus, now manufactured by a medical company, turned out to be valuable in a wide variety of cases involving respiratory problems.

Nearly forty Pulsators were known to be in use by 1939 in hospitals in Birmingham, Ipswich, Liverpool, London, Manchester, and Norwich, and they were put to the test during an epidemic of poliomyelitis in 1938.[33] There was also an unusual number of diphtheria cases in children, and as the Pulsator was portable, it was particularly valuable in emergency situations. One testimony to its efficacy came from Ireland, where a consultant at the Cork Street Fever Hospital in Dublin was fighting a high incidence of diphtheria.[34] He wrote to Bragg:

> I never saved a case before I got the Bragg-Paul Pulsator, and on behalf of 38 of my patients whose lives *you* have saved I thank you most profoundly.

The same doctor was also able to respond to an emergency appeal from Northern Ireland, when the superintendent of Belfast City Fever Hospital requested the loan of a Pulsator, which was duly dispatched in time to save a child's life. It would have continued to be widely used and modified

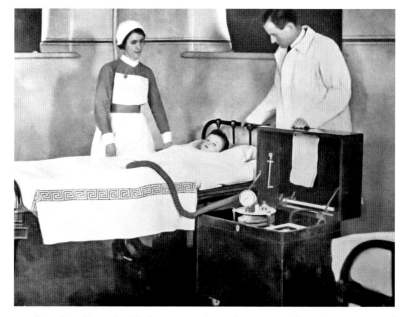

74. Over forty Bragg-Paul Pulsators were in use by 1939, mainly in the treatment of diphtheria and polio, and the device's successful use at the Cork Street Fever Hospital in Dublin was documented.

had not Lord Nuffield, the founder of Morris Motors, offered to supply an alternative, even cheaper respirator to hospitals throughout Britain and the empire at his own expense. The last improvisation in Paul's long career of invention brought him what must have been gratifying recognition in an unexpected field—in fact, a film illustrating its uses was made at the University College London Physiology Department in 1939.[35]

CHAPTER 11

Paul and Early Film History

In May 1921 the sixty-five-year-old William Friese-Greene attended a meeting of film distributors in London, called to discuss the postwar crisis facing British producers. After an impassioned speech urging unity among members of the British cinema industry, he collapsed and died. Friese-Greene had long claimed that his patent no. 10131 of June 1889 for a moving-picture camera made him the true inventor of moving pictures, bringing a series of lawsuits against those he believed had ignored his pioneer status.[1] One of these—against the Kinemacolor system of Charles Urban and G. A. Smith—had led to the collapse of Urban's business in 1914.[2] Nonetheless, Friese-Greene's destitution at the time of his melodramatic death, together with a fulsome and highly inaccurate tribute to him by Will Day, launched a tradition that reached its climax in *The Magic Box*, a film starring Robert Donat, made in 1951 as a collective memorial by the British film industry in 1951, for the Festival of Britain.[3]

Although Friese-Greene's claims to priority have been disputed, he remains Britain's only film pioneer commemorated with such ceremony, and his elaborate grave in Highgate Cemetery (designed by Sir Edward Lutyens) still bears the inscription "the inventor of kinematography." His death in 1921 also helped inspire the creation of a "Cinema Veterans" association.[4] A "veteran" was defined as one who was "actively involved in the Cinema Industry in (or before) 1903, and remained therein for a reasonable period."[5] One hundred and one people who met this criterion apparently attended the first annual dinner in December 1924. Many of these would have been Paul's colleagues in the early 1900s, and although it is uncertain whether Paul was present, he had become treasurer of the association by 1929.[6]

Paul's interest in organizations that contributed to public awareness of science and technology is much better documented and seems to

75. William Friese-Greene (1855–1921; *left*), photographer and serial inventor, failed to profit from his achievements and died in abject poverty. Long considered in Britain to be the "true" pioneer of moving pictures, on the basis of an 1889 patent application, his reputation peaked with a sentimental 1948 biography that became the basis of *The Magic Box*. Though his achievements were challenged in the 1960s, later research has absolved him of claims made after his death, showing that he did in fact create moving pictures in 1891. Will Day (1873–1936; *right*) had many roles in the early decades of British cinema, primarily as an exhibitor and dealer in film and radio equipment. His collection of historic artifacts was displayed at the Regent Street Polytechnic in 1936, and for several decades at the South Kensington Science Museum, before being bought by the Cinémathèque Française in 1959. Day bore much responsibility for misleading claims made after Friese-Greene's death in 1921.

have intensified in the 1920s. He had been a member of the Institution of Electrical Engineers since his student days and is recorded as playing an active part in its debates in 1919.[7] He became a fellow of the Physical Society of London in 1920 and served as a vice-president between 1928 and 1930.[8] From 1927 he was also a founder and fellow of the new Institute of Physics, which would later incorporate the Physical Society, also serving as one of its vice-presidents.

In 1921 Paul formally joined the Royal Institution, where he had originally demonstrated the Theatrograph in 1896, and he served on various of its committees during the 1930s, also donating examples of his equipment. Founded in 1799, and invariably headed by a leading scientist since

76. A major exhibition at the Royal Albert Hall in 1931 celebrated the centenary of Michael Faraday's discovery of electromagnetic induction, launching "the science of electrical engineering and the great electrical industry as we know it today." For this, Paul collected examples and built working replicas of over two hundred measuring instruments, many of which were demonstrated in operation.

Sir Humphry Davy became director of its laboratory in 1801, the Royal Institution established a tradition of offering accessible and often spectacular lectures, at a time when science was still poorly represented in British universities. Under Davy's leadership, Michael Faraday became first a lecturer, then director of the laboratory, where he carried out his groundbreaking experiments in electromagnetism. Faraday also contributed a record nineteen series of the RI's annual Christmas Lectures, aimed at popularizing aspects of science; the lectures were first televised in 1936, as part of the BBC's experimental television service.[9] Paul contributed filmed visual aids to Sylvanus Thompson's Christmas Lectures series in either 1896 or 1912.

One of the RI's major projects, which deeply engaged Paul, was a Faraday centenary exhibition, jointly promoted with the Institution of Electrical Engineers and staged at the Albert Hall in September 1931. Paul's *Times* obituary reports that "he took a very active part in preparing the collection of scientific apparatus" for this exhibition. However, as another obituary notes, "Probably no voluntary worker did nearly as much for the Faraday Exhibition as did Paul; but in the long list of acknowledgements, the name of Paul does not appear."[10]

It is not difficult to see why Paul should have been enthusiastic about commemorating Faraday, since so much of his own career had followed from Faraday's momentous experiments and discoveries. The opening statement of the exhibition catalogue eloquently and precisely explains his importance:

> In 1831 Michael Faraday . . . in his Laboratory at the Royal Institution . . . made the discovery in which lies the origin of the dynamo and starting point of the utilisation of electric power for the purposes of man. On that day . . . he wound two coils of wire on to opposite sides of a soft iron ring, connected one coil to a battery and the other to a galvanometer, and at "make" and "break" of the battery circuit observed deflections of the galvanometer. . . . From this simple experiment . . . have grown in the past hundred years the science of electrical engineering and the great electrical industry in all its phases as we know it to-day.[11]

The exhibition layout in the central arena of the Albert Hall was intended to form a kind of map of what had flowed from Faraday's pioneering work.[12] Radiating out from his statue, surrounded by cases containing "apparatus that he made and used, including the famous ring of 29 August, 1831, and the diary in which for forty years he entered day by day the results of his work," twenty-five sections showed what had followed from the primary experiments, in chemical engineering and the distribution of electric power, and on to telegraphs and telephones, broadcasting, and domestic uses of electricity."

The idea was that visitors would, as they proceeded along the aisles, "observe the movement . . . of the marvellous applications of electricity to modern life" and "follow the lines of reasoning that have carried it so far."[13] Paul's contribution to this impressive educational project becomes clear only from close reading of an article he contributed retrospectively to the *Journal of Scientific Instruments*.[14] His main interest, unsurprisingly, was in measurement, and in making these specialized devices understandable to a lay public, in what was a pioneering example of what would today be called an interactive, hands-on exhibition:

> In all, more than thirty instruments were shown in action, either by the staff of honorary demonstrators, or on the operation of a switch by the visitor.[15]

Paul's practical involvement began when "a preliminary survey of the subject showed a paucity of original apparatus." He discovered through "reference to original publications . . . a wealth of detail . . . of the apparatus and its principal dimensions," which made it possible "to construct a number of reproductions." Paul appears to have been responsible for many, if not most, of these re-creations of the generation of instruments that had preceded those he pioneered. He wrote:

> Prior to the 1880s, electrical instrument making was not recognised as a distinct industry, only becoming such after the birth of electrical engineering, [before which they] were made by firms known as philosophical instrument makers, and usually constructed throughout by individual workmen of great skill.

He then traced the impact of his own teachers, Ayrton and Perry, who, following Marcel Duprez, introduced instruments graduated in degrees "which could be converted to amperes or volts," adding by way of an inside joke:

> To them is due the abbreviation "ammeter" now in general use for amperemeter, but strongly opposed by Sylvanus Thompson, who sarcastically asked "Why not 'vometer' also?"

Paul's survey of the development of electrical instruments from the earliest times to the near-present feels scrupulously objective, never referring to his own role in this progress. Yet, knowing what we do about his pride in and commitment to the business, it feels also like a very modern form of "show and tell," with the promise that the instruments loaned and specially manufactured would be on display at the Science Museum, "possibly improved and amplified," for an extended period in the following year.

Always practical, Paul went on to endow an Apprentices Prize at the Physical Society, and later the Instrument Fund that still bears his name, to promote scientific instrument design and innovation.[16] His achievements in science were recognized by the award in 1938 of the Institute of Physics' Duddell Medal, a prize given biannually for distinguished contributions to the application of physics in an industrial, commercial, or business context. As an indicator of its standing, the previous winner was a German, Hans Geiger, co-inventor of the Geiger counter, and the next was the American nuclear scientist Ernest Lawrence, creator of the cyclotron.

Becoming a Part of Early Cinema History

Paul had been actively involved in the step-change in instrument making that took place around 1900, and had obviously undertaken research for his contribution to the Faraday exhibition, which marked an important step toward establishing the history of modern technology.[17] By the early 1920s, not only were cinema "veterans" coming together socially, but the first steps were under way to organize a history of the unruly film business. In 1920 Terry Ramsaye, then a journalist writing for *Photoplay* magazine, had published a long series of articles surveying the rise of the motion picture. Realizing the potential for a more substantial and detailed history, he set about corresponding with as many as possible of the pioneers—one of whom was Robert Paul. As already noted, *A Million and One Nights* (1926) included a chapter boldly titled "Paul and the Time Machine," which made Paul an inescapable part of the story that, in Ramsaye's ebullient telling, led from Aristotle and Leonardo through the optical discoveries of the eighteenth and nineteenth centuries, to the launch of Edison and Dickson's Kinetoscope.

At which point the story repeated at the outset of this work, of Paul's commission to replicate the Kinetoscope, made its second published appearance, after Talbot's 1912 account in *Moving Pictures*.[18] Ramsaye quoted a letter from Paul, written on 23 July 1924, about his contract with Acres, dated to 29 March 1895, which now became a part of the record.[19] Ramsaye also took the opportunity to dismiss the "enthusiastic patriotic claims" made by Talbot (who was not named) for Friese-Greene, which had "caused a great deal of complication and misunderstanding in motion picture affairs."[20] By contrast, he concluded, "Paul, conservative and precise, does not engage in undocumented memories."

Ramsaye clearly also saw an opportunity to capitalize on Paul's "time machine" patent, which he claimed had been "utterly forgotten" until his researches brought it to light. On one hand, it allowed him to bring H. G. Wells, by now a world-famous figure and author of a global *Outline of History*, into the long prehistory of moving pictures. But the links Ramsaye strenuously worked to make between the experience of the Kinetoscope and Wells's passages describing the appearance of time travel also tied into the overarching theme of his book:

The Wells-Paul idea, embodied in the patent application, contained gropings for a greater and new liberty for art. It sought to liberate the spectator from the instant of Now. . . .

77. H. G. Wells (1866–1946) shot to worldwide fame as the author of a series of "scientific romances" that followed his breakthrough with *The Time Machine* in 1895. His early biological education earned him wide respect among scientists, while his *Outline of History* (1920) became an influential book during the interwar years. Although his ideas permeate much early cinema, and clearly film influenced his early fiction, Wells had little success in writing directly for the screen until Alexander Korda produced the spectacular *Things to Come* (1936) from his script.

> This motion-picture Time-machine idea was artistically at one with Ouspensky's mathematically mysterious philosophy and Einstein's philosophically mysterious mathematics. It was a promise of a more concrete application of their remote intellectual abstractions.[21]

At that time, Wells claimed he was "unable to remember details of the relation" with Paul, although once reminded, he would include it in the introduction to his 1929 book *The King Who Was a King*, which recounted an unsuccessful attempt to script a fictional film.[22]

Typically, Paul appears to have said nothing further on being reminded of his farsighted patent of 1895. But he had remained in contact with fellow veterans, especially Will Day, and in 1932 was prompted to recall his early days in moving pictures. The occasion was a long letter from Oskar Messter, a German film pioneer of the same generation, who was trying "to clear up [details of] the commencement of Kinematography in Germany. . . . Therefore it is also necessary to clear up the commencements in other countries."[23] Messter reminded Paul that they had met, "at the dinner at the Grand Hotel in Paris," during the European conference in 1909. Now he wanted to put some precise questions:

1. Since what time do you sell kinematographic projectors which worked professionally with success and of what construction were these?

2. Were these apparatuses already manufactured by series?

3. How many of these apparatuses were sold directly till the end of 1896 and practically used?

Messter asked also for a copy of Paul's advertisement showing a series of projector models from 1895 to 1901, adding, "In the text, so far as I remember, that the apparatus was not successful practically before 1897." And he returned to the issue of relations with Acres:

From Hopwood I see that Mr Byrd [*sic*] Acres had taken kinematographic photos already in March 1895. Were these shown in public with success and were projectors which could be used practically sold in England before your apparatus was published [i.e., marketed]?

Paul's reply was cordial: "I well remember the Conference in Paris, when you and I were appointed to the committees of our respective countries," adding, "I hope that we may meet again, and if you come to England you will visit me at your convenience."[24] His answers covered what had become familiar ground, no doubt benefiting from the research done to answer Ramsaye.

1. My first sale of a projector was in March 1896, to David Devant; it resulted from a descriptive article in the *English Mechanic* on March 6: Mr Devant purchased immediately a machine and exhibited with it at the Egyptian Hall.... The films used were those I had already taken for the Kinetoscope.

2. I made two only of this type, in which an oscillating shutter was employed. The first had been exhibited in February 1896. Of the type shown in your sketch, without continuous feed for the film, and with two intermittent sprockets, the total number did not, I believe, exceed 24. This type was suited for short films only, and was very quickly replaced by a Pro-

jector with "top feed," as shown in my Patent Specification 4686, dated March 2. These I made in batches of 24 or 36 at one time.

In answer to Messter's third question, Paul said that he no longer had the business records of that period, when "persons came from many distant countries . . . often offering large premiums for precedence in delivery," adding that "these offers were never accepted." He mentioned Méliès and Charles Pathé as "early buyers" but gave only the global figures used in the flotation advertisement (see appendix B).

> I estimate that the number of projectors sold in the year was between 200 and 300. All were practically used and many buyers made large sums by exhibiting in various countries, including the Far East.

Paul rejected any suggestion that "the apparatus was not successful practically before 1897" and said of a nonfunctioning projector reported in Berlin, "Either it bore my name, cast into the iron base-plate or it was not of my manufacture." "I was aware," he added, "that copies were being made, in a crude fashion, by at least two mechanics and, on account of the strong demand, sold in a few cases at high prices."

The familiar story of Georgiades and Trajidis and the Kinetoscopes followed, up to the point of Edison refusing "to supply buyers of my machines with films." On his dealings with Acres, he wrote:

> In January 1895 I became acquainted with Mr Birt Acres, a travelling representative for Messrs Elliott and Son, makers of plates and paper for photography, of Barnet. He informed me that Mr Blair of Foots Cray, Kent, was making photographic film and could cut it to the required width for the kinetoscope. Neither of us had any experience of "Cinematography," but Mr Acres was a photographer. He proposed, and it was agreed, that I should make a perforator, camera and printer, which he would use to make pictures exclusively for me. On March 29 1895 a formal agreement was signed to that effect. I had already completed the apparatus and installed the printer in my factory; the first film had been taken outside the house of Mr Acres, and a cutting from the print then made is enclosed. About 20 subjects were taken, including the Boatrace [*sic*] and Derby.

Later in the letter, he added, unequivocally, "With regard to Mr Birt Acres's early work, I am sure that he had no experiments in moving pictures before he had the camera which I designed and made in 1895."

On the "time machine," Paul only remarked, tantalizingly, that "Mr Wells called on me, by request, to discuss possibilities and he made some suggestions as to the scenes." Messter had asked about the mechanism used to create intermittent motion, and Paul explained that his first machines used a seven-point sprocket (half of the Kinetoscope's fourteen-point sprocket, "which had evidently too much inertia for an intermittent action"), "modified to four-point and three-point in my subsequent projectors." He commented, in answer to Messter's main inquiry about the four-point Maltese Cross mechanism:

> It is interesting to see the survival, in some many projectors, of this type of motion. I am pleased to see that you adopted it at so early a stage in the development of projectors, and with so great success.

Paul also added, alluding to his work for the Faraday exhibition, "It is a pleasure to answer your questions . . . having recently spent much time on finding or reproducing the early instruments for electrical measurement."

Four subsequent letters from Paul to Messter survive (though apparently none of Messter's), up to 1935, offering information that Paul had been able to get from correspondents. An issue of concern to Messter had been claims that the Skladanowsky brothers may have anticipated his own success in 1896; on this Paul was able to reassure him:

> One of my friends who has long occupied himself with the history of cinematography tells me that Max Skladanowsky was invited to substantiate to the Royal Photographic Society in London his claims to be an early inventor of a camera and projector, and that he was unable to produce satisfactory evidence. Further, he doubts if any wide film was available as early as November 1895 and suggests that Skladanowsky be asked who was the maker of his film.[25]

Paul's responses must have helped Messter in writing his autobiography, published in 1936, in which he was happy to admit that his own first projector was based on a copy of Paul's Theatrograph.[26] Although this was not the case when Messter first wrote, by 1935 the issue of who was first to project film in Germany had become political, after the Nazis came to power in 1933. Long forgotten, the Skladanowsky brothers were resurrected and feted, on the basis of their having shown films in November 1895, as Stephen Barber has explained:

Determined to recast all cultural lineages in order to highlight the preeminent role of Germany, the Nazi cultural authorities seized on the Wintergarten Ballroom film-projection event [of November 1895] in order to elevate the now-elderly and semi-forgotten Max Skladanowsky to the status of a great German inventor. The propaganda minister Josef Goebbels appeared at a gala-evening dedicated to Max Skladanowsky at the lavish Atrium cinema in Berlin on 4 May 1933, and Hitler himself attended a private screening of the Skladanowsky Brothers' films in 1935.[27]

It seems unlikely that Paul would have known about these events, or that Messter would have mentioned them specifically, amid the more obvious reasons for concern over Hitler's Germany at this time. But as 1936 dawned, celebration of the beginnings of film was already on the agenda for Britain.

Commemorating 1896

A British Kinematograph Society had been launched in 1931, after separating from its parent, the American Society of Motion Picture Engineers, primarily to organize training initiatives for different branches of the film industry.[28] This it did initially through the London Polytechnic in Regent Street, which had been offering its own courses in cinematography since 1908, now reinforced by formal connections with the industry. Exactly who initiated or coordinated the celebrations of the fortieth anniversary of moving pictures is unclear, but two main events took place under the auspices of the BKS and of the Polytechnic in the anniversary month of February. First, the BKS held a meeting on 3 February with the title "Before 1910: Kinematograph Experiences," with three of the most important of surviving pioneers, Robert Paul, Cecil Hepworth, and Will Barker, giving papers that were subsequently published.[29] Then, on 20 February, a ceremony was held at the Polytechnic to open an exhibition of historic equipment, at which Louis Lumière was the guest of honor.

The BKS symposium was the only occasion on which Paul spoke before an audience about his career in film, and the text, cited throughout this study, has become one of the foundational documents for understanding his career. From the instances of reminiscence already cited, it seems clear that Paul became aware that he had a unique testimony to give, and despite writing to Messter in 1935 that he was "rather averse to personal publicity," he was persuaded to condense his memories into a formal speech—certainly more formal than those of his colleagues. Both of these acknowledged what they owed to Paul's pioneering work. Hepworth had literally got his start in film through Paul, being first inspired by seeing a "sideshow of the Theatrograph" at Olympia, when he went there with his father, the famous lanternist. He then managed to sell six of the hand-fed arc lights he had created to Paul, before buying some of Paul's cast-off fragments to mount his first film shows. Barker had chaired the Kinematograph Manufacturers Association, and had worked on topicals with Paul's general manager, Jack Smith.

Paul's own paper was divided into ten headed sections, tracing a chronological progression from "the Kinetoscope era," through "Projection by Intermittent Motion" and "Selling Projectors," to "Events in 1897," which covered Victoria's Diamond Jubilee and the disastrous Bazar de la Charité fire. The Jubilee had established for Paul the importance of covering topical events, and getting the films out to exhibitors quickly, while the Paris fire drove him to develop a fireproof projector. But he admitted that "after the Jubilee public interest in animated pictures seemed to be on the wane." Foreseeing the need for "long films giving complete stories," he began the development of the Muswell Hill site, which would house the studio these required. He summarized two of his approaches to the Boer War: sending cameras to the battlefields, which produced "about a dozen good films," and, "to meet the demand for something more exciting," making careful "representations" of typical scenes, which he also admitted were probably passed off as the real thing by showmen.

From references in the text to "a portion of a film" and to slides, it is clear that Paul's presentation was extensively illustrated.[30] He showed images from two early trick films, one, probably *Diving for Treasure* (1900), seeming to show undersea divers at work, achieved by filming figures in real diving suits against a back cloth, with "a large narrow tank, containing live fishes between the stage and the camera." Paul then commented revealingly:

> The result appeared sufficiently natural to cause the Prince of Wales and Lord Rothschild, after seeing it on the Alhambra screen, to ask me how it had been possible to photograph under water.[31]

Paul's screenings at the Alhambra were long finished by 1900, so one or another detail of this story would seem to be wrong. As with most of Paul's recorded claims, there is no source of independent verification. But the remark seems to confirm that film had catapulted him into very different social circles from the early electrical engineering community.

The other film shown, *A Railway Collision*, which "was considered very thrilling" and widely pirated in America, had been made using small model railway rolling stock, as revealed by a slide of "the setting in the garden." After describing the organization of his studio and processing plant, Paul paid tribute to "Walter Booth and others" before analyzing in some detail the effects used in *The Magic Sword*, "as an example of what was done at the beginning of this century to pack the maximum of movement into 180 feet of film."[32] Interestingly, he coupled the scientific demonstration films he made for Boys, Worthington, and Thompson with other "trick films," implying that these were all aspects of film's capacity to conjure the otherwise invisible.

The final sections of the talk, "Improving the Projector" and "Selling Films," covered the two branches of his film business in the new century, noting that both had continued to progress, and indeed thrive. Jack Smith's work in "topicals" and the *Army Life* series were recalled with particular pride. And the end of Paul's career in cinema was foreshadowed by condemning the tendency among "less conscientious showmen" to make commercial use of films they had been sent for approval, and by the rise of competitive price-cutting around 1907. Paul paid tribute to Will Barker for having "called the producers together" to form the KMA, and modestly hoped that their fellow speaker, Cecil Hepworth, would "tell us about the European Conference in 1909."[33] Last came his now-familiar explanation of his reasons for abandoning film, with the elegant *envoi*:

I look back with pleasure on the childhood of your art, and with gratitude for many personal kindnesses received and much loyal service rendered.[34]

Paul Kimberley's cryptic reference at the end of the meeting to having known Paul "in the days which he termed 'Reformation'" has already been mentioned. The seconder of the vote of thanks, a Mr. Whitehouse, referred to "the element of enterprise and real courage [all three] had to advance this new thing," and ended by invoking broadcasting as the latest "great machine of entertainment."[35]

There was, however, a more public dimension to the 1936 London anniversary celebrations. On 20 February, Louis Lumière was the guest of honor at the Regent Street Polytechnic, for a "reproduction of the original programme of forty years ago," compèred by Cecil Hepworth and projected by the collector and chronicler of cinema's beginnings Will Day. Some additional "early silent films" were also shown, "by the courtesy of the British Film Institute," an organization that had been founded

CINEMA VETERANS
who unite in homage
to
LOUIS LUMIÈRE

CECIL HEPWORTH
One of the pioneers of Kinematography. Started production in 1896. Supplied pictures to America in early days, and responsible for some of earliest and best British pictures.

R. W. PAUL
Showed pictures at the Alhambra, March, 1896. First manufacturer of Projector for commercial use in this country. Erected first Film Studio in U.K.

78. The fortieth anniversary of cinema was marked at the Regent Street Polytechnic in June 1936, with Louis Lumière as guest of honor. Paul and Hepworth were featured among other British veterans acknowledged, and some of Paul's early equipment was shown in an exhibition organized by Will Day.

in 1932 despite determined opposition from within the film industry. Extracts from "later silent films," "the talking picture," and "colour films" made up the rest of the evening's program, shown in the same hall that had originally housed the Lumière screening of 20 February 1896, now operating as a regular cinema.[36] The evening's program brochure included two pages of "Cinema Veterans who unite in homage to Louis Lumière," with photographs of twelve. Paul appears on the top right, with the caption "Showed pictures at the Alhambra, March 1896. First manufacturer of Projectors for commercial use in this country. Erected first Film Studio in this country." Alongside him, Hepworth is described as "One of the pioneers of Kinematography," with his start in production backdated to 1896, and praised as "responsible for some of the earliest and best British pictures."[37] According to an account of the evening, Hepworth had jokingly claimed that no one had had a greater influence on his life and career than the Lumière brothers: "We prayed for light, and God gave us . . . Lumière."[38]

This evening served as the opening ceremony for an exhibition, in the nearby Fyvie Hall of the Polytechnic, of more than twenty items from Day's pioneering collection — in effect the forerunner of all subsequent chronological "story of cinema" displays, and no doubt a reminder for Paul of what he had made possible for the Faraday exhibition five years earlier.[39] This exhibition started with "moving lantern slides of 100 years ago," and various optical toys and devices of the nineteenth century, up to Reynaud's Praxinoscope. The film-related exhibits began with "Edison's Original Kinetoscope," accurately described as "the machine which first gave Mons. Lumière the incentive to show moving pictures upon a screen in France, also Mr R. W. Paul in England." What followed, however, reflected the belief Day had helped foster, that "The First Film Made and Patented in the World" was an unperforated strip, the "Invention of W. Friese Greene, Bristol, whose Patent no 10131, was the prior patent for a Motion picture . . . date June 1889." Another strip, described as "The Second Film," now perforated, was dated 1890. Four pieces of Lumière equipment were on display, with the Cinématographe unnamed but described as "The First Mechanism." "The First Theatrograph Projector," attributed to Paul in February 1896, carried the additional information, "taken by Carl Hertz on an African tour," while two Edison items were strangely identified as a "Biograph Projector" and an "Edisonograph" projector.

As usual, there is no way of knowing how closely Paul was associated with either the "tribute" event or the exhibition — or what he might have thought of the narrative that was beginning to take shape, in which Lumière had been granted so central a role. We know virtually nothing about what continuing interest he may have taken in cinema. Would he, for instance, have gone to see the film that premiered at the Polytechnic Cinema on the day after the tribute, Alexander Korda's futuristic spectacle *Things to Come*, based on the latest of his onetime collaborator H. G. Wells's cautionary fictions?[40] Leslie Wood claimed to have interviewed Paul on the eve of the Second World War, and said Paul had recalled the "time machine" project "with a twinkle in his eye," saying:

> We soon realised that we would have to spend thousands of pounds on machinery and a suitable auditorium, so it did not get any further than being a project on paper.[41]

79. The last known photograph of Paul. Recognized by a Norwegian filmmaker on the quayside, while his car was being unloaded from a Fred Olsen ship, Paul asked for a copy of this photograph and promised to send a film he had made of the last occasion HMS *Victory* sailed off Portsmouth (probably his 1903 film of *Victory* in collision with the *Neptune*). The film never arrived. Photo from Jan Anders Diesen, *Filmeventyret beginner: av og filmpioneren Ottar Gladtvet* (Oslo: Norsk Filminstitutts skriftserie 10, 1999).

The only other report of Paul's later views on cinema comes from the apprentice mechanic who worked with him during the interwar years:

> One of the rare occasions when Paul was in a talkative mood, he said "I did not think cinema would take on in the way it has; and as for the money these stars get! I got my artists from the Music Halls, they were very glad to get some daytime work. One day we wanted a washerwoman for a set and found what was needed just down the road."[42]

At the end of 1936, Paul's BKS paper was republished in the American *Journal of the Society of Motion Picture Engineers*, having been "requested and recommended for publication by the Historical Committee." The text is substantially the same as its published English version, although without the references to Paul's contemporaries, and with fourteen illustrations, many of his equipment. There are, however, a number of significant minor differences. In the later version, Paul wrote that he pointed out to Acres . . . the intermittent motion in our first camera was not suited to give accurate spacing, and the pictures compared unfavourably with the original Edison films. So soon as the camera was put into use, I therefore proceeded to make a second with which many of our 1895 films were taken.[43]

A footnote in this version records that their *Rough Sea at Dover* "was included in the program shown at Koster and Bial's Music Hall, New York, on April 23, 1896, when the Armat Vitascope was used to project the pictures." This in turn is referenced, presumably by the editors, to an article by Armat in the same journal the year before. In the following year, 1937, Paul was again approached to confirm details of his early film work. A young South African scholar, Thelma Gutsche, was embarking on a doctoral thesis, "The History and Social Significance of Motion Pictures in South Africa," and must have asked him for information about the Anglo-Boer War films.[44] His reply confirmed that the second camera he sent to the Cape was "in [the] charge of Mr Sidney Melsom, who (with my two brothers) was in the CIV."[45] He then listed the more successful of Colonel Beevor's films, adding that, although he was happy to provide the information, "I fail to see what interest these particulars can have at this date."

This rather curt response to one of a new generation of film scholars was Paul's last contribution to film history, although not to the saga of the "time machine." Thereafter it would be the business of others, who had had no direct contact with him. But the onset of another world war brought new demands, albeit not on his practical scientific skills. As a senior director of Cambridge Scientific Instruments, he had to weather various government attempts to bring about mergers between various specialist companies in this field.[46] Above all, with two key board members both seriously ill in 1941, Paul had to chair an important meeting shortly before Christmas.[47] He had been pressing the board to bring in "new young blood" with scientific expertise, and his handling of the meeting produced a confrontation with the new managing director, William Apthorpe, who had become manager of the Muswell Hill premises after the 1919 merger. Paul was sufficiently concerned to write to Robert Whipple, the chairman, and his old friend, on Christmas Day:

I think you will understand my concern about the future of the business, a small part of it being the outcome of many years of my own labour. I regard my present position as anomalous but . . . I am more worried about what I think is a vulnerable position ill-adapted to cope with post-war difficulties.[48]

Early in the new year, Whipple was well enough to resume being an active chair, and apparently worked to improve relations between Paul and the other directors. In 1944 he donated his own large collection of historical scientific instruments to Cambridge University, creating what would become the Whipple Museum, at the center of the university's Department of History and Philosophy of Science. Paul, meanwhile, had died fifteen months after the boardroom confrontation, on 28 March 1943.

The cause of death was recorded as cerebral thrombosis, with chronic bronchitis and emphysema also contributing (no doubt exacerbated by the heavy smoking habit his assistant had noted). His health must have been deteriorating, as he died in the Priory hospital at Roehampton, while Ellen was living nearby, in Wimbledon, presumably to facilitate visiting him in hospital. Paul's will left her an allowance, with his shares in Cambridge Scientific Instruments being used to found a trust, the R. W. Paul Instrument Fund, that would make awards to further design and construction of apparatus to measure phenomena in the physical sciences. The fund still operates and is administered by the Royal Society.[49] Two years after his death, Ellen applied for an increase in her allowance, arguing successfully that she had not been left sufficient funds under the will. She lived until July 1954, remaining at Victoria Drive.

Paul's grave in Putney Vale Cemetery bears two unusual motifs, decipherable only by those with an intimate knowledge of his marriage and his "original business." The raised motif in bronze on the stone is a schematic version of the unipivot instrument, which he — or perhaps Ellen — regarded as his greatest achievement. But the phrase "true till death" surely refers to Irene Codd's story of the Pauls' courtship, when Robert gave Ellen this challenge. Whatever difficulties Ellen may have experienced in Robert's last years, or over the provision of his will, she clearly kept that promise.

Paul's Afterlife

When Robert Paul died, in the middle of the Second World War, most obituaries paid tribute to his work in scientific instrument making, with only a paragraph about his "sideline" in film, following the phrase that Paul himself had used. However, there would soon be a new impulse to probe what had initially been a sideline for most of the pioneers. The project of carrying out research on "the authentic history of the British cinema both as an art and as an industry" was formulated by the British Film Institute in 1946, as a way of marking the fiftieth anniversary of the first film screenings. It was to be steered by a Research Committee chaired by Cecil Hepworth and including another veteran filmmaker, George Pearson, along with Ernest Lindgren, curator of the National Film Library, the forerunner of Britain's National Film Archive. A young researcher, Rachael Low, was employed to work alongside the lecturer and writer Roger Manvell on a first instalment covering the period 1896–1906, which appeared in 1948.[50] This slender volume of just 136 pages became the only guide to the first decade of film in Britain, with an appendix cautiously listing "First Recorded and Claimed Demonstrations of Motion Pictures Projected on a Screen," which was noted to be an "extremely controversial" subject. What made it so was the credence still given to Friese-Greene's claims— or those of his supporters, as already noted—soon to reach its climax in *The Magic Box.*

The biography on which that film was based, *Friese-Greene: Close Up of an Inventor,* appeared in 1948, written by a young Ulsterwoman, Muriel Forth, under the pseudonym "Ray Allister." Forth was apparently inspired to take up Friese-Greene's cause after seeing his funeral in 1921 and becoming aware of the campaign to establish him as the "true" inventor of cinematography. She was not, however, a reliable researcher, and her amateurish book followed what Terry Ramsaye described in a dismissive review as "the cliché pattern so dear to Sunday supplement journalism."[51] Although Forth's book was well enough received in Britain to become the basis of *The Magic Box,* it was greeted with scorn in the United States, not least because it claimed that Friese-Greene's 1889 patent effectively invalidated Edison's. In the longer term, it almost certainly led to Friese-Greene's actual achievements being underestimated by those intent on demolishing the myth of his priority.[52] And in the short term, it completed the transfer of the "policeman" story, first misattributed by Day in 1921, from its rightful source with Paul to Friese-Greene.

As already noted, Paul himself told the story at a KMA dinner in 1908, and Leslie Wood correctly attributed it in his 1937 popular history, telling of a policemen in Hatton Garden hearing

> excited cries emanating from Robert W. Paul, aided and abetted by his assistants, roaring out war-whoops of joy.... The constable was annoyed at this commotion and, to appease him, Paul showed him the marvel which he had just discovered.[53]

80. *The Magic Box* (1951). Muriel Forth's sentimental and inaccurate biography of William Friese-Greene became the basis of an equally misleading film, made as the British film industry's contribution to the Festival of Britain celebrations. Robert Donat's ecstatic achievement of showing moving pictures supposedly taken in Hyde Park to a bemused policeman was a fantasy based on Robert Paul's actual celebration of his and Birt Acres's first film being shown on a Kinetoscope in 1895.

In *The Magic Box*, it is Robert Donat as Friese-Greene who shows his first success to an initially skeptical policeman, played by Lawrence Olivier. What is elided—both in Wood's reference to the "excited cries" and in Forth's book, as well as in *The Magic Box*—is the crucial gap between movement only visible through a Kinetoscope viewer, which Paul and Acres achieved in March 1895, and movement projected onto a screen, which Paul did not achieve until early 1896. Although the film's historical plausibility was widely attacked, especially in America, neither of these confusions appears to have been remarked.

Low and Manvell's *History*, however, is also far from reliable on what Paul had achieved up to 1906. Organized analytically in two parts, "Production, Distribution and Exhibition" and "The Films Themselves," it begins by identifying Hepworth and Paul as "the two most important British producers" of the first decades, and cautiously establishes the latter's Olympia show in March 1896 as

the first performance in this country of motion pictures projected on to a screen, at which an admission fee was charged, given by an Englishman.

He also seems to have been the first to sell cinematograph apparatus and films.[54]

Thereafter, there is a surprising amount of misinformation, compounded by a misprint that has Paul retiring in 1900 (instead of 1910) and only subsequently achieving distinction in the scientific world.[55] The output of his "famous Muswell Hill studio" is described as "not large — some fifty films a year," when in fact Paul was by far the most prolific early English producer, with over eighty titles in 1896 and 1897, rising to over a hundred in 1900. But the most damning statement is Low's dismissive comparison with Hepworth:

> His importance to the British film industry ... was due rather to his business and scientific abilities than to any artistic gifts. Led into film production by force of circumstance, he remained an engineer rather than an artist, and his films seem, from a perusal of their detailed synopses, to have been characterised by a lack of either taste or appreciation of the larger possibilities of cinema, which contrasts strongly with the films of the Hepworth Manufacturing Company.[56]

The fact that Hepworth was alive during the preparation of Low's first three volumes may have influenced the respectful treatment given to his career.[57] But probably more significant was an aesthetic position apparent throughout, shared by Ernest Lindgren's contemporary *Art of the Film*, which set store by early signs of "advanced technique."[58] Thus Williamson's early railway scenes were implausibly described as exploiting "the cinematic qualities of trains, anticipating films which ... fifty years later would make magnificent use of the same material, such as *Nightmail, La Bete humaine, Brief Encounter.*"[59] By contrast, Paul's pioneering *Army Life* series was said to have "made no attempt to deal with any men individually as characters" — an observation that seems willfully to ignore the stated aim of this overview of the Army's many branches.[60]

Low's treatment of the period was largely, if not entirely, based on catalogue synopses. Although she recognized Paul as having produced the first "made-up" (in Hepworth's term from the 1936 symposium) or story film, with *The Soldier's Courtship* of 1896, she quotes the catalogue description of his 1898 remake. Continuing the comparison of Paul and Hepworth, she writes:

> Paul's films are interesting both for their lack of subtlety and comparative absence of the more advanced technique to be found in Hepworth's later

films made towards 1906 ... the comedies are rather comic scenes or situations than comic stories. The technique of a comic twist to a story does not appear in his work.[61]

Paul was also criticized for dealing with "the dubious comedy of the domestic dilemma type leading to strife in the home," leading Low to suggest that "it may be necessary to question Paul's own taste in this matter," while "the other chief directors of this period (such as Hepworth, Williamson, Smith, Haggar) do not seem to have made so much game out of marital strife, drunkenness, absence of clothes, physical cruelty and other themes recurrent Paul's work." Above all, Low sought to demonstrate how the "series picture" (using another Hepworth term) evolved from "magic lantern technique" to "organic continuity from one [shot] to another," holding up Hepworth's 1905 *Rescued by Rover* as the earliest example of "modern film technique with editing playing an important part in the design and movement of the film."[62] And indeed this film, the last discussed in the book, is the only one apparently described from viewing.[63]

Low's suggestion that the plots of Paul's dramatic films might reflect "lack of taste," or even some moral laxity on their producer's part, may seem bizarre today. Yet it would prove influential, as all subsequent writers have read Low's account of Paul in the course of writing their own — despite its being based on little or no actual viewing. Most surprising is the same reference to Paul's "lack of taste" by the French historian Georges Sadoul, whose *Histoire Générale du Cinéma* first made the claim that early filmmaking in Britain was more advanced than in other countries, including France and America.[64] Sadoul's ten-page chapter "The English Cinema before 1900" placed Paul at the center of an account that stressed how the profusion (*eparpillement*) of British producers in the early years was a source of originality and commercial success, before it became "archaic after 1902, when cinema became industrialised and concentrated."[65] Paul, characterized as "a merchant of optical and scientific apparatus," was oddly presented as part of an English tradition of magic lantern makers who had realized the necessity of providing their customers with "a varied and extensive repertoire of views." Yet his 1898 illustrated catalogue was described as "more imposing and better presented than that of any English or French contemporary," as well as being "bilingual," which assured the films "many purchasers in France."

Most striking of all, however, was Sadoul's close analysis of Paul's first steps in multiscene narrative in 1898, which he described as more "evolved" that those of Lumière "or even Méliès" at that time. He singled out *Our New General Servant* on the basis of its catalogue description of sequential, spatially distinct scenes, and also *The Stockbroker*, before citing the scenario of *The Deserter* as an example of a direction that "other producers in other countries would still follow ten years later":

> perhaps the first example of a "social" realism in British cinema, whose international influence would be seen especially on Pathé and Vitagraph and later Griffith.[66]

Sadoul judged Paul's comedies of the same period as less original, while admitting that the use of two settings in *Come Along, Do!* was "in this ultra-primitive era, yet another innovation which detached cinema from theatre." His account of Paul's studio, built to enable more complex productions, noted that it had "relatively more complicated machinery than Méliès had at Montreuil at the same time." Striking an oddly anachronistic note, Sadoul characterizes the open-air stage Paul pictured in his catalogue as "closer to a large cupboard than our modern studios," but adds the intriguing information that when A. C. Bromhead started to build a studio for Gaumont in Britain, inspired by Paul's example, Leon Gaumont told him to "stop this extravagance," so that the studio was not built until 1902.

Sadoul's chapter on early British film gave due recognition to British Mutoscope and Biograph, and to the "Brighton school" of Collings, Smith, and Williamson, which he was the first to celebrate and cite as the source of a "real artistic originality" in the British industry after 1900. Yet examples of Paul's originality and ambition dominate his account, with the shrewd comparative observation that the Transvaal war played the same role for British film as had the Cuban war for America and the Dreyfus affair for France, in each instance giving rise to "reconstituted actualities." In Paul's case, he distinguished the many genres Paul used to address the Anglo-Boer War, leading up to the public and press recognition he received for the *Army Life* series. But after noting his advertisement of November 1900 proclaiming "1,000 subjects: comic, patriotic, trick, war and domestic," he concluded that the resumption of peacetime genres was "so primitive" that Paul could not repeat his earlier success.[67]

Further evidence that Sadoul was impressed by his close reading of Paul's catalogues came in an unexpected reference in a May 1948 review of *Lady in the Lake*, directed by and starring Robert Montgomery as detective Philip Marlowe. After describing the film's unusual "first person singular" strategy, in which the camera represents the protagonist's point of view such that we see only what Marlowe is supposed to see, Sadoul wrote:

> The principle of this process is not new. In 1900 William Paul released a little film taken from a car travelling through Piccadilly Circus, narrowly avoiding collisions with buses and cabs. Already, the camera had "become an actor." The spectator, who imagined himself a passenger, trembled as he trembled when, in the Grand Café, he thought the locomotive at La Ciotat advancing towards him, was going to kill him.

Quite apart from revealing that the mythology of spectators terrified by the Lumières' *Arrival of a Train* was already established, this otherwise contemporary review shows that close reading of film catalogues, even when the films themselves were not available to view, could persuade unbiased historians of their originality.

Sadoul had clearly studied Paul's catalogues more carefully than any historian before, or indeed since. He was particularly alert to the frequent borrowings by Ferdinand Zecca at Pathé, described as "a regular plagiarist of the English optician [*sic*] and in this respect something like his pupil."[68] He also noted a number of likely imitations of Méliès in Paul's 1901 trick films, "which owe much to Montreuil and are not original except in their vulgarity." This charge echoes Low's criticism, with Sadoul claiming that "the vulgarity of Paul can be shocking," a condition he attributed to the influence of Paul's "fairground customers."[69] It remains a baffling judgment, especially when repeated from a French perspective, but it did not prevent Sadoul from noting how much the leading French filmmakers of the era owed to Paul: his *The Gambler's Fate; or, the Road to Ruin* (1901) is identified as the "original" of Zecca's *La vie d'un joueur* (1903), "one of Pathé's first great successes." Perhaps most surprising, Sadoul concluded that Paul's *Adventurous Voyage of the Arctic* (1903) must have inspired Méliès's last major film, *Le voyage au pôle* (1912), a full nine years later.

For and against Paul

During the next thirty years, the early history of film in Britain became a minor province of world film history, with a few anecdotes and a general sense of anticlimax substituting for any systematic treatment or new research. Charles Oakley's review of "seventy years of the British Film In-

81. *The Gambler's Fate; or, The Road to Ruin* (Paul, 1901; *top*). (b) *La vie d'un joueur* (Ferdinand Zecca, 1903; *bottom*). Georges Sadoul notes in volume 2 of his *Histoire générale du cinema* (1948) that this and a number of Zecca's other early films were closely modeled on Paul's earlier titles.

dustry," *Where We Came In*, appeared in 1962, and its cursory treatment of Paul was typical. Oakley's coverage of the "Prince's Derby" was compressed, so that the Prince of Wales saw his horse win on-screen on the evening of the day the race was run. Paul's connection with the Alhambra was extended to four years. The success of *The Soldier's Courtship* is credited with leading to the idea of the New Southgate studio. And Paul's total output of films was said to be eighty-two. Oakley seems to have borrowed Low's verdict, that Paul "had little talent for directing" and that his films were "commonplace." His final departure from production in 1910 is attributed to the failure of an expensive trick film, *The Butterfly*—which even now remains untraced.[70]

The late 1970s saw the beginning of a new wave of empirical research

82. *The Conquest of the Pole* (1912). Georges Sadoul believed that Méliès's last film, with its ice giant puppet figure, showed a clear influence of Paul's 1903 *Adventurous Voyage of the Arctic*.

into early film history in Britain. Two important publications preceded the now-legendary International Film Archives conference in Brighton in 1978, at which an unprecedented range of pre-1906 films was actually screened.[71] In 1976 John Barnes published *The Beginnings of the Cinema in England*, chronicling the years 1894–1896 in scrupulous detail.[72] For Barnes:

> The dominant figure in the history of the cinema in England was undoubtedly Robert W. Paul, or "Daddy" Paul, as he was later affectionately referred to in the trade, because above everyone else, he can be truly regarded as the father of the British film industry.[73]

In response, the following year saw the first in a series of articles by Richard Brown, of which the second, "England's First Film Shows" set out to question Paul's status and to assert the importance of his temporary associate, Birt Acres:

> It will be suggested that previous accounts dealing with Paul's involvement in [early] cinematography ... need modification, and that many important details in Paul's own accounts are either untrue or intentionally

misleading. The role of Birt Acres will be assessed, and his earliest demonstrations of projection established for the first time. It will be concluded that it was from the inventive skills of Acres, that Paul derived much of the knowledge which later enabled him to rise in importance in the early English film industry.[74]

Apart from casting doubt on Paul's claims to have played the leading part in creating what Barnes called the "Paul-Acres" camera, and shedding new light on how Paul may have progressed from his Finsbury College demonstration to the Royal Institution and on to the Alhambra, Brown's articles asserted that "effective pioneering actions were stimulated by a business environment rather than being the result of a purely aesthetic or technological bias."

Barnes and Brown would continue to take opposing sides on the Paul-Acres issue, and in doing so probably stimulated a wider continuing engagement with the material at issue.[75] But Brown's emphasis on the importance of the "business environment" introduced a new complexity and sophistication into what had long been largely a matter of anecdote or mere assertion. The next phase of early cinema research would introduce further social and political perspectives, as well as formal analysis. Michael Chanan's 1980 study of early film in Britain, *The Dream That Kicks*, located it within continuing traditions of popular entertainment, noting Paul's close connections with such key figures in modernizing stage magic as David Devant and Nevil Maskelyne, and also the new opportunities for commercial exploitation of invention.[76] He cited Paul's own telling observation in 1936 that the number of associated patents registered between 1890 and 1895 had been 63, while the next ten years saw this mushroom to 566. In his major empirically based study of the evolution of film style, Barry Salt identified Paul as one of the earliest users of a number of key innovations, from joining single shots in 1898 (*Come Along, Do!*) to undercranking for accelerated motion onscreen (*On a Runaway Motorcar*, 1899) and a textual insert shot in *Buy Your Own Cherries* (1904).[77]

Noel Burch brought to this new wave of interest in early film both a filmmaker's interest in formal structures and a sharply political reading of its subjects. In *Life to Those Shadows*, he challenged Sadoul's claim that Paul, along with Williamson, Haggar, and Mottershaw, "generally sides with the humble," locating him instead "in the ideological space of a completely conformist middle class."[78] Singling out Paul's *His Only Pair* (1902), in which children outside a window jeer at a poor child while his trousers are mended ("an example of a quite obscene miserabilism"), and *Goaded to Anarchy* (1905), where "the revolutionaries are conveniently far away, in Russia," he claimed there is a "striking contrast both in 'language' and manifest content" between the populism of early French film and the English production of "entertainment artefacts made by bourgeois manufacturers for the urban working classes."[79]

Several short books, intended to summarize what was by now understood as the complex field of early cinema, included valuable perspectives on Paul. Emmanuelle Toulet's *Cinématographe, invention du siècle*, appearing in a series that prioritized illustration, offered a succinct overview of the "small production companies in Britain that developed a quality cinema," giving priority to Paul, but ending with this shrewd observation:

> English production, despite an exceptional attention to narrative and a strong influence on [other] filmmakers of the period, played only a modest part in the international market for film. More than its stylistic innovations, it was its typically British spirit which limited the export of English film.[80]

A short textbook, Simon Popple and Joe Kember's *Early Cinema: From Factory Gate to Dream Factory*, placed Paul in the context of a discussion of "technological determinism," drawing attention to his early offers of equipment to the Science Museum as part of a tradition of understanding cinema through its evolving technology. Popple and Kember also cite the "antagonistic debate" conducted between Acres and Paul during 1896 in published letters as an instance of the "personalisation" of early cinema history, whereby the protagonists sought to "establish personal recognition for their contribution . . . to safeguard their legal status and to exploit the patent right to cinematographic equipment."[81] Perhaps the most scrupulous and balanced account of the disputes between Acres and Paul, and of their achievements in the early years, appeared in Deac Rossell's *Living Pictures*, which drew on new research from Germany supporting Acres's position, while also showing how internationally influential Paul's Theatrograph had been, in both Germany and France.[82] Acknowledging the many currents that shaped early film studies, Rossell avoids partisanship in his account of the 1890s, making the important point that, whatever the contributions of the many individual inventors and suppliers, it was the "brand" of Edison that dominated the early spread of moving pictures—however little he may have contributed personally to their achievement.

Luke McKernan, a historian who has researched the career of Paul's contemporary Charles Urban, wrote in a brief online biography of Paul for the British Film Institute:

> Paul's career seems to have been blissfully free from crisis; he simply did everything right, and left the film industry at exactly the right time. Although a filmmaker himself for a short period only, his creative and industrial influence was immense.

This may be unexceptionable, but it seems rather bland. Does it not underestimate the extraordinary bursts of invention shown by Paul in 1898 and again in 1901, as well as the commercial risks he must have taken to maintain his film business after the failure of the attempted flotation in 1897? Not to mention the international influence on other filmmakers traced by Sadoul? And here we must also wonder about the organization of Paul's studio: how much autonomy did he give his employees, even if he appears to have been nothing like as remote from actual production as Edison or Charles Pathé?

Questions of Attribution

That Paul employed a number of staff in his studio has always been known and acknowledged, with at least six or eight of these going on to their own notable careers in the British film industry.[83] But one of his collaborators, Walter Booth, has increasingly been singled out as the "director" of a significant number of Paul's fictional and trick-based films. This attribution seems to have been started by Dennis Gifford, in his *British Film Catalogue* (1973), possibly influenced by his having traced and contacted Booth's descendants, and runs through subsequent editions of this influential catalogue. It now appears as an unquestioned fact in almost all references to such widely discussed films as *Upside Down, Scrooge,* and *The ? Motorist,* with Paul being credited, if at all, as "producer."[84] John Barnes took up this challenge in the final volume of *Beginnings of the Cinema in England*:

> We cannot be sure who the directors of these films [of 1900] were. Walter Booth may have responsible for some and R. W. Paul for others; some films may have been co-directed. Unlike the films of Georges Méliès, they do not bear the unmistakable style of the artist. If anything, they have the hallmark of Paul himself, a rather methodical down-to-earth character with the precise and innovative flair of the scientist, who was certainly not averse to team work.[85]

Twenty years later, John Barnes's brother and collaborator, William Barnes, would write in the introduction to a program of Paul's "magic films":

> One of John's last articles before his death in 2008 . . . was entitled "The Quest for Walter Booth." Even without access to our newly found portraits of Booth . . . John was convinced that Booth was the principal performer in the Paul films. He pointed out however that though Booth might have arrived at Paul's studio well equipped with the tricks of the stage magician, the magic in the Paul films is entirely produced by a skilled film-maker, using the resources of the camera and the laboratory and a repertory of stop-motion, double exposure, camera inversion, laboratory manipulation.[86]

The fact that we now have a means of identifying Booth as a key performer in Paul's films should, if anything, raise doubts about his role as "director."

However, the more fundamental questions remain unanswered, and perhaps unanswerable. What basis is there for positing the role of "director" in the period when Booth worked for Paul, since the word was certainly not used, nor did the role, as we understand it today, exist until considerably later? And as the Barnes brothers have asked insistently, what evidence did Gifford have for his claims, apart from the assumption that creating "film magic" would be easier for a practicing magician, albeit a stage magician?

Epilogue

What are the implications of claiming that Paul was one of the most active and engaged of the "pioneer" film generation in Britain, or indeed anywhere in the world? Such a claim is hardly fashionable today. It uncomfortably recalls the "auteurism" that first launched film studies in the 1960s, before coming under attack as a reactionary methodology, ultimately supporting a "great man/artist" ideology. Since then, there has been a succession of approaches to the complex communications revolution of the late nineteenth century, with increasing attention rightly paid to networks rather than individuals, and to business as much as to art.

In many ways, the recovery of a large cache of films by the Blackburn company Mitchell & Kenyon in the late 1990s has provided a template for the "new history" of early film, directing attention toward social context and away from the metropolitan scene, and from the claims of "firstness" and innovation that once preoccupied historians of early film.[1] In line with prevailing approaches, a recent *Companion to British Cinema History* refers only briefly (and inaccurately) to Paul as being "in business" with Birt Acres in 1895–1896. So is Paul doomed to have been marginalized once at the beginning of "British film history," and again seventy years later—an inconvenient pioneer?

And what if we redraw the map of early cinema, to make London equal to Paris or New York as one of the hubs—or, to change metaphor, incubators—of the moving-image industry and its culture? We know from Paul's own account that customers converged on his workshop in 1896 to buy Theatrograph projectors, and undoubtedly films as well. But Paul was only alone for a short time. Just as there were other British Kinetoscope manufacturers in 1895, so the number of equipment suppliers and producers of films would grow rapidly. Simon Brown has shown in his study of British cinema's original base in Cecil Court, near Leicester Square, that "agglomeration" was an inevitable dynamic driving the emergence of a "new medium" industry.[2]

In fact, London had been in the forefront of new forms of spectacle ever since Philippe De Loutherbourg first showed "moving pictures" in his Eidophusikon just north of Leicester Square, in 1781. Twelve years later, Robert Barker opened the world's first purpose-built panorama building, also on Leicester Square, and throughout the nineteenth century, London was renowned for spectacular displays of panoramas, dioramas, and "dissolving views," as well as becoming an international center for the modernized magic lantern, and for stereoscopy and other new visual media.[3] It also witnessed, near the end of the century, a rising tempo of experiments in motion photography, as Paul himself noted, which included Wordsworth Donisthorpe's Kinesigraph as well as the overstated results of Friese-Green's patent 10131, both in 1890.[4] The immense success of the Kinetoscope in Britain during 1894–1895, as documented by Brown and Anthony, was not confined to London. But the capital provided a concentrated and ready market, already avid for new entertainment experiences, which would support the wave of manufacturing and services of which Paul became a leading part.

Researching and writing this book has necessarily involved collecting as much biographical and business material as possible, as well as constructing and exploring Paul's filmography of some eight hundred productions — around the same number as produced by Mitchell & Kenyon, although fewer than eighty of Paul's films are currently known to survive.[5] Like the "missing" films, the figures of Paul as technologist, entrepreneur, employer, and husband have all needed to be reconstructed. The result almost inevitably looks like a celebration, a claim that Paul deserves to be better appreciated — and indeed I do make this claim, largely because so little attention has hitherto been paid to gathering the most basic information about his foundational career(s). Only in Britain, with its deep-seated, class-based prejudices toward cinema, and indeed toward business and technology, would such achievement be so little remembered or commemorated.

But if Paul deserves some measure of overdue recognition, he also cries out for adequate interpretation and contextualization, as a leading figure among Britain's new media pioneers — a figure in the tradition of De Loutherbourg, Sir David Brewster, or Sir Hubert von Herkomer, combining skills and sensibility in timely new ways. From his own self-presentation, we have inherited the impression of film as a "sideline" to the more serious business of research in the new electrical industries.

This is clearly how Paul saw himself, and apparently wanted to be seen, at the end of his life. No claim to be the "father of the British film industry" on his tombstone, in stark contrast to the hyperbolic praise heaped on Friese-Greene after his death. In fact, we know nothing about Paul's opinion of that industry and its works after 1910 — except that he understandably considered it too "speculative" to remain involved in. Instead, we find him being honored by scientists and technologists, and becoming fascinated by the history of electrical science.

In that paper he gave in 1936, carefully judged and crafted, he admitted to an important stake in film "before 1910." But what does his final phrase, "the childhood of your art," imply? Certainly distance, but what were his deeper feelings about the seventeen years he spent juggling film and instrument making? If we accept that he was managerially and creatively involved in film between 1894 and 1910 — at times almost certainly more deeply involved with film than with supervising the instrument business — then did he not play a significant part in creating the vast audience for what would become cinema? Rachael Low's history has cast a long shadow, even though attitudes toward, and access to, early film have changed dramatically. The engineering skills that underpinned both of Paul's businesses brought him esteem in one but perversely damaged his reputation in the other. To devise and sell film equipment was hardly seen as contributing to the "young art and educational force" that had "forced its way to a position of cultural importance," as Low proclaimed at the beginning of her history.[6]

Paul can be regarded as an early casualty of the distinctively British antagonism between science and culture that erupted in the 1960s as the "two cultures" spat between the scientist C. P. Snow and literary critic F. R. Leavis.[7] Equally, he can be seen as a victim of Britain's uneasiness about cinema itself, with its assumed vulgarity — a long-running complaint that also owed something to Leavis's antipopulist stance in the 1930s,[8] and has regularly resurfaced, as recently as 2017 in an exchange between the English novelist and sometime film reviewer Adam Mars Jones and director Martin Scorsese.[9] Trying to piece together the components of Paul's dual careers has led me to the conclusion that they informed each other, with experimentation and "product optimization" prominent in both practices. The Paul who continued to improve his projector and camera over many years was also the producer who strove to make his films more "gripping," who added titles and tinting to make them more attractive and comprehensible, who organized them in series that anticipated the feature-length of later productions by as much as a decade. As the main British producer for at least five years, he should be credited with an important role in turning mere "animated photography" into cinema. And above all, with his many innovations, both technical and artistic, helping to create its mass audience.

Engaging Audiences

Traditionally, the early audiences for moving pictures are seen as generic: mere pleasure-seeking masses who patronized the music halls and fairgrounds, about whom we can know little as individuals. The earliest sociological attempts to probe audience motivations and attitudes date from the decade after Paul left the business. But "his" decade produced a surprising number of imaginative responses to the new experiences film was offering. These range from Gorky's story "Revenge" (July 1896) and George Sims's "Our Detective Story" (1898) to Maurice Normand' s "Devant le Cinématographe" (1900), Kipling's "Mrs Bathurst" (1904), and Andrei Bely's essay "The City" (1907).[10] It may be coincidental that two of these involve films by Paul, but it is his early successes of 1895 and 1896 that give us some insight into what attracted early spectators. The first screening of his Derby film at the Alhambra in June 1896 is one of the few documented occasions in the early history of exhibition, with reviews telling us what music was played and that the film was repeated by popular demand. Screenings in Cheltenham and Brighton that same year also yield valuable contextual information about how and to whom film was being presented.

As Paul consolidated his production in the following years, we can trace a series of distinctly experimental strategies, aimed at attracting new and more varied spectators. First came the "tour" series in 1896– 1897 (Spain and Portugal, Egypt, Sweden). His systematic coverage of Queen Victoria's Diamond Jubilee in 1897, although not unique, provided an "ideal" composite form of spectatorship. As already noted, Paul launched his 1898 season with what amounted to a manifesto, aimed at buyers, but focused on the "telling effect" these films would have on the "attention of beholders"—almost certainly the earliest reference by any producer to audiences.[11] After the outbreak of the South African war at the end of 1899, Paul responded to the level of public interest with a remarkable range of films in different genres and experiments with multiscene dramas, from topical records to allegorical tableaux and battlefield "reproductions." His Army Life series was also the earliest attempt, apart from the first records of boxing matches, to create a complete screen-based event on a single theme.[12] A new range of "attractions" followed in

1901. Paul attempted to link film with performance in "animated songs," to be shown with specially commissioned live music, ahead of Gaumont launching its Chronophone synchronization of film and disk in the following year. His *The Countryman and the Cinematograph* (1901) provided a template for many subsequent films that found humor in the cinema experience itself, mocking naïve spectators.[13] For the end of 1901, in the season traditionally dominated by festive entertainment, he produced ambitious trick-based "remediations" of Dickens and of pantomime, in *The Haunted Curiosity Shop* and *The Magic Sword*, two of the most technically advanced films being made anywhere at that time.

Although the turn of the century appears to have marked the peak of Paul's filmic innovation, his catalogues reveal a continuing commitment to offering new forms of audience engagement. In 1904–1905, major foreign events were dramatized; and for documentary series, the "tour" and the "visit" were supplemented by the "day in the life" format, applied to coal mining (*A Collier's Life*, 1904) and racehorse training (*The Life of a Racehorse*, 1904). Reports in the trade press of a reorganization of Paul's studio facilities in 1908 may indicate that he was changing the balance between his film and instrument businesses. But until the dramatic events of 1908–1909, he seems to have remained optimistic and fully engaged in the emerging film industry, with documentary-style films of increasing scale, such as *Whaling Afloat and Ashore* and *The Newspaper World from Within* (1909), the latter running for nearly twenty minutes. Still, the realization that exhibitors and renters could dictate terms to producers in the British market, and could also benefit from cheaper and more exploitable imports, must have shaken his confidence—at least as much as the formation of Edison's cartel.

Electrical Parallels

In this climate, consolidating and expanding his electrical business must have seemed an attractive option, and 1909–1911 seems to have been when he took enlarged showroom premises in Leicester Square, as well as opening an American branch of the instrument company. Behind these bare facts, however, lies an extraordinarily complex history of how electrical power became central to the modern world between 1890 and the First World War—a history to which Paul belonged, as the explosion of generating systems and new uses of electricity demanded a growing range of precision measuring equipment.[14] This is also a history that is even less widely known than that of early moving pictures. The names of Marconi and, of course, the tireless self-publicist Edison may be familiar, although the latter is better known internationally for the phonograph and moving pictures than as an electrical pioneer. But Faraday, Paul's hero, and William Thompson, later Lord Kelvin, are probably remote grandees to many today, in part due to the legacy of Britain's "two cultures" tending to downgrade science and technology. We have already heard from Hugo Hirst, Paul's contemporary and the founder of the General Electric Company, and about his teachers Ayrton and Mather. But how many could identify the contributions of Sebastian Ferranti, R. E. B. Crompton, the Siemens brothers, Charles Parsons, or Charles Mertz and his partner William McLellan, all key figures in making electricity universally and economically available?

Gavin Weightman's popular history of "how electricity changed Britain forever," *Children of Light*, brought together brief biographies of these and many other colorful entrepreneurs and visionaries who eventually brought order to the vast turn-of-the-century mosaic of small companies producing electric power for local use.[15] What emerges from Weightman's account is a pattern bearing some similarity to the consolidation of moving pictures, with early scientific achievements stifled by commercial confusion, ideological hostility, and, in the case of electrical generation, misguided legislation. The specialized branch of the industry that occupied Paul, devising and manufacturing instruments to meet new monitoring and educational needs, must have been affected by some of these industrial pressures. And his two "industrial" speeches that we have, from 1914 and 1919, deserve to be understood in a wider industrial and political context.[16] Film historians risk underestimating the larger uncertainties involved in the electrical industry, and also perhaps failing to explore parallels between the film industry that Paul was central to, and his involvement with another new industry, both local and global.

New Approaches to Early Film

By a strange irony, just three years after Paul left the film industry, another unlikely figure declared his faith in "the possibility of art in the film," suggesting that "the black and white artist never had such a chance as now, with the cinema by his side."[17] Sir Hubert von Herkomer was a distinguished German-born painter and graphic artist, a Royal Academician who had already created an art school and private theater at his estate in Bushey, before he announced in 1912 that he was embarking on film production.[18] Herkomer's new venture at the age of sixty was cut short by his

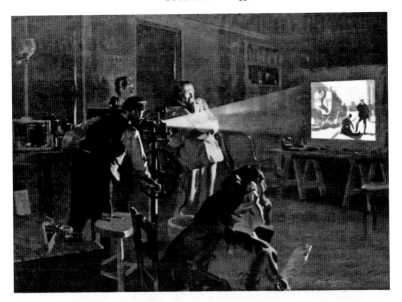

IN THE HATTON GARDEN WORKSHOP of a scientific instrument maker, in
1895, strange new pictures were thrown on a magic lantern screen —
pictures that *moved*. What may have seemed merely an
intriguing novelty to the admiring witnesses was, in fact,
the first commercially practicable film projector to be made
in this country — the Theatrograph.
Its inventor was Robert W. Paul, one of the purposeful
men who made the 1890's a period of promise unique in our history.

Another was Albert E. Reed, who that same year began to make
super-calendered newsprint and other printing papers at Tovil,
near Maidstone, having converted an almost derelict straw mill
acquired the previous year. Expanding his paper-making
business with the energy and enterprise so typical of the times,
he founded one of the largest paper-making organisations in
the world. For to-day the four mills of the Reed Paper Group,
with its unrivalled technical experience and resources, produce

Pioneers in modern paper technology

Reed

83. A 1952 advertisement in *Punch* for the Reed Paper Group drew a parallel between
Paul's throwing "strange new pictures on a magic lantern screen" in a Hatton Garden
workshop in 1895, and the start of Reed's papermaking the same year. The inspiration
for the ad is unclear, although it partly corrected what was attributed to Friese-Greene
in the previous year's *The Magic Box*—while maintaining that film was *projected* in
Britain in 1895.

death after only a year's filmmaking, but it serves to remind us that there
was a wider culture developing around film before 1914, which did not see
it solely in commercial terms.

In her discussion of "the case of Herkomer," the art historian Lynda
Nead has asserted that "it makes no sense to write the history of any single
medium," which can only "deliver a pre-determined set of responses," as
in such cases as "the failure of nineteenth-century British painting, or . . .

the early years of film."[19] She argues instead for "an integrated history of visual media," especially at moments when "new and old visual forms were at their most tenuous and unformed, and when visual culture and new technology were meeting head-on." Nead explores the relations between Herkomer's late move into film and his early passion for motoring—a conjunction that could shed light on Paul's similar interests outside his main career trajectory.[20]

There is no equivalent manifesto on the artistic potential of film from Paul, although his 1898 claim that "the capacity of animated pictures for producing BREATHLESS SENSATION, LAUGHTER and TEARS has hardly been realised" might be considered one of the earliest forecasts of what cinema would deliver (see chapter 7). What this account of Paul's moving-picture career has tried to do is place not only his output of individual films, but his approach to creating a new repertoire and a new modern experience, within the culture of which he was a part. If we compare his situation with Herkomer's, Paul was already engaged in the quintessentially new electrical industry when he embarked on creating animated photographs that would "rivet the attention of the beholders." He was, in effect, remediating much of the previous century's popular narrative culture. Nearly fifteen years later, Herkomer declared his intention of abandoning painting in order to create a "nerve-resting," rather than "nerve-racking" cinema, one "which does not necessarily consist of constant thrills and excitements and sensations."[21]

Herkomer was speaking nearly twenty years after the first display of moving pictures as a novelty, when "cinema" had become an established zone of social life, with its own national and international traditions, and had become predominantly narrative. Paul's most active period belonged to the era "before the nickelodeon," to recall Charles Musser's title for his study of the near-contemporary Edwin Porter, Edison's most successful director. Paul's films always formed part of programs that he soon had little or no control over: the epitome of the "cinema of attractions," as first defined by Tom Gunning and André Gaudreault and taken up by many others following in their wake.[22] That such programs should contain varied attractions was firmly believed by film professionals of the era, as Theodore Brown's "ideal" programs of 1907 made clear.[23] The correlative of such programs was the audience they attracted, as described by the Russian poet and early film enthusiast Andrei Bely in the same year:

The cinema is a club. People come together here to undergo a moral experience, to travel to America, to learn about tobacco farming and the stupidity of policemen, to sigh over the *midinette* who has to sell her body. Absolutely everyone comes here to meet their friends: aristocrats and democrats. Soldiers, students, workers, schoolgirls, poets and prostitutes.[24]

However, such an account should not simply be taken at face value, as Tsivian made clear when quoting it. The Muscovite Bely was arguing against an aesthetic view of cinema held by the Petersburg Symbolists, painting cinema as "utopia next door," a down-to-earth form of togetherness, which became the template for how cinemagoing would be characterized for the next decade.[25]

Paul's surviving films, amounting to around one-tenth of his total output, are now more accessible than they have ever been, thanks to the internet. Viewers are free to watch, to download, and even to annotate and playfully add to them. They have, in a way, escaped into a very modern kind of digital utopia, unconfined by the copyright structures that have limited access to the contemporary films of Lumière and Méliès. Generally, they are seen by innocent modern viewers as charming, quirky, quaint—genuinely fragments of the childhood of *our* era, and perhaps of our moving image arts.

The idea of considering them as single visual texts, specimens of early "film form," now seems anachronistic, a throwback to the first epoch of discovering early film in the 1980s. Increasingly we feel obliged to see them as a part of visual and popular culture, and also as "intermedial," enmeshed in a variety of media and cultural "memes" that provided their subjects and framed their early circulation. As the progenitor of a body of work, encompassing the films, catalogues, reviews, and related historical incidents, Paul appears as a unique figure in early cinema, perhaps comparable only to Porter or to Méliès, but exceeding these in significant ways. Across the period 1895–1909, he engaged in every issue that faced early film as an industry, and indeed as an art. He worked in every genre and format known during this period, even if much of the evidence of this is only indirectly available to us—and his deployment of color is completely lost, as are his allegorical tableaux and "illustrated songs." Yet for all that is missing, perhaps most valuable today is the *discourse* that frames his work, allowing us to follow Paul's reading of the evolving field of production for the film market.

In biographical terms, Robert Paul's swerve into a demanding new role as showman of "animated photography" is more difficult to understand than his judgment about leaving the industry, although his "invisible" manner of quitting remains hard to credit—except as a major example of the impatient decisiveness he showed in other fields. There was an anecdote about him smashing a botched piece of work by one of his employees, urging him to make a fresh start, and this may be the best explanation we will ever find as to why he abandoned film. Having seen the directions that cinema was starting to take in 1909, he knew that to participate, to compete commercially, in this new era would require an investment of resources that could only be at the expense of his electrical instrument work. In short, he had to choose.

Time and Space Travel

Does any of this make Paul a "time traveler," beyond that unconsummated liaison with Wells in 1895? We have no way of knowing whether he continued to read Wells, or whether Wells ever saw any of his films after the Kinetoscope period. If Paul attended the Polytechnic event of 20 February 1936, which included examples of his own early work in film, he could hardly have avoided noting that the Korda film of Wells's *Things to Come* was being premiered on the following evening. But would he have been tempted to attend? If he did, he may well have been as disappointed as Wells was by this schematic, overly didactic prophecy of a war that nearly ends civilization, which drew on his superior "dream diary" novel *The Shape of Things to Come*. This and other film projects suggest that Wells never really grasped how effective films are structured — except perhaps with three short comedies that he outlined in 1928, to help his son Frank, recently started in art direction, get a foothold in film production. In the unlikely event that Paul saw these absurdist silent comedies, he might have been reminded of some of the "comics" he had produced twenty years earlier, with their stock "policeman," "burglar," and "tramp" characters.

However, we now know that at some point in their later lives, Paul and Wells did meet, thanks to a letter from Paul to the scientific journal *Nature*, in response to an article by Wells in 1941. With a typical flourish, Wells had claimed that Paul's 1895 "time machine" patent was a joint application "that would have made us practically the ground landlords of the entire film industry."[26] Paul was quick to respond, with a letter in which he amplified earlier accounts of the contact with Wells:

> I invited Mr Wells to my office in Hatton Garden towards the end of 1895. He listened patiently to my proposals, gave his general approval to my attempting to carry them out, and proceeded to talk of subjects suitable for

the primeval scenes. Having recommended for my perusal some books on extinct monsters he left without further discussion as to future action.[27]

To which Paul added, tantalizingly, "Many years elapsed before I had the pleasure of meeting him again." But whether he saw *Things to Come*, or followed Wells's increasingly pessimistic predictions about the danger to humanity from technological warfare, remains unknown.

Wells's literary reputation had faded by the time of his death in 1946. But it began to rise in the 1960s, bolstered by Bernard Bergonzi's claims for the early scientific romances as "mythopoeic explorations" of key Victorian preoccupations such as evolution and class war.[28] Might similar perspectives be usefully applied to Paul's oeuvre, insofar as we can "read" it across the catalogues and surviving fragments, thus countering the 1940s accusations of "vulgarity" by recognizing Paul's subjects as eminently and robustly typical of their Victorian and Edwardian era. Given prevailing attitudes to early film as "primitive" in the 1930s and 1940s, it is hardly surprising that Paul should have viewed his film career as infantile from the viewpoint of 1936. But today's "media archaeology" rejects such teleological judgments and calls for accepting early media devices and texts *on their own terms*. What we find in Paul, from the combination of surviving films and catalogue descriptions, is an unusually viewer-centered process, aimed in two main directions. One of these can best be understood in terms of "remediation": making film tokens of such literary classics as *The Last Days of Pompeii* and *A Christmas Carol*, as well as versions of contemporary stage successes (*Two AM, Trilby*). The other direction seems to be driven by an ambition as great as any of Paul's contemporaries—Albert Kahn, for example, with his Archives of the Planet—to realize the "territorial" potential of film, mapping Britain, as well as half a dozen other countries of topical or exotic interest to British viewers.[29]

By the end of his film career, Paul's catalogue offered a gazetteer of subjects likely to engage domestic audiences, ranging across Europe, Scandinavia, and the empire. What had started as, effectively, a "time machine" in the tradition of Edison's Phonograph, had become a new kind of "space machine," enabling spectators to visit an expanding global repertoire of sites and spectacles in a new way.[30] Today's cultural geography joins media archaeology in reminding us that the decade when Paul was most intensively involved in realizing the potential of "animated photography" was when the intrinsic interconnection of space and time was first proclaimed.[31] Thirty years later, the Russian cultural theorist Mikhail Bakhtin would borrow the term "chronotope," which he had found both

84. *Fun on the Clothesline* (1897). Recently discovered in the BFI National Archive, this simple comedy features the slack-wire performer Harry Lamore, together with what appear to be Robert and Ellen Paul, filmed in a garden setting.

in Einstein and in biology, to express the "intrinsic connectedness of temporal and spatial relationships that are artistically expressed":

> The special meaning it has in relativity theory is not important for our purposes; we are borrowing it for literary criticism almost as a metaphor (almost, but not entirely). What counts for us is the fact that it expresses the inseparability of space and time (time as the fourth dimension of space).[32]

Bakhtin's concern was exclusively literature, although he recognized that this new concept might have wider applications, wherever "spatial and temporal indicators are fused into one carefully thought-out, concrete whole." "Time," he continued, "becomes artistically visible; likewise, space becomes charged and responsive to the movements of time, plot and history."

It is tempting to suggest that the cartography of Paul's reconstituted film catalogue demonstrates this connectedness of space and time, as well as that of any other filmmaker of the pre–First World War era. And indeed new discoveries and connections continue to appear. During the time that this book has been in active preparation, four previously lost films have been identified, one of which has been successfully restored from fragments (*A Collier's Life*).[33] Fuller understanding of the lifelong link between Paul and H. G. Wells has also emerged, from film's presence in *The Sleeper Wakes* to their 1941 exchange in *Nature*. And even further back, new research on the mysterious visitors who propositioned Paul in late 1894 has revealed that Georgiades and Trajidis were just two of a group of entrepreneurial Greek emigrants who may have been responsible for introducing the Kinetoscope to Europe.[34] Ahead of Edison's licensed agents, they exhibited the device in Paris as early as July 1894, and it seems likely that their Paris shows inspired Antoine Lumière to urge his sons to develop what became the Cinématographe. While one of their associates continued to manage the Paris show, Georgiades and Trajidis saw new opportunities in London, which led to their introduction to Paul. While the Greek entrepreneurs were prosecuted for "passing off" Paul's machines as Edison products, it appears they may also have urged Paul to find a way of making new subjects to supply their—and by then his own—Kinetoscopes. This is speculative, but it seems possible that the development of the Lumière Cinématographe and Paul's Theatrograph have a common point of inception in Greco-American entrepreneurship in the early 1890s—foreshadowing the immigrant generation of Adolph Zukor, Samuel Goldwyn, and Louis Mayer as founders of the great Hollywood studios.

More light has been shed on another technical aspect of the Paul-Acres collaboration by Peter Domankiewicz, who has resumed research on William Friese-Greene. At the British Silents symposium at Kings College, London, in April 2019, Domankiewicz suggested that Paul and Acres succeeded in producing their camera so quickly in early 1895 thanks to Acres having seen Friese-Greene's descriptions of his camera in the photographic press in 1890—a hypothesis that deserves further attention. Moreover, he demonstrated convincingly that the "incident outside Clovelly Cottage" test film must have been shot at what would become projection speed, rather than the normally higher Kinetoscope rate.

More evidence has also appeared of Paul's scientific connections at the turn of the century. The Royal Society has a record of him providing, and very probably operating, a projector for a soirée on 20 June 1900, when A. C. Haddon showed his now-famous film of the Torres Straits islanders' initiation ceremony.[35] Linking this with the fragmentary information about his other scientific work summarized in chapter 10, it would seem that Paul's presence in both scientific circles and the emerging world of film was known and valued.

Yet, as we have seen, Paul remained modest about his "sideline." And another recent archival discovery transports us back to the "childhood" of film, very much as he must have experienced it. Bryony Dixon, curator of silent film at the National Film Archive, has discovered an unsuspected early Paul film in the collection, which seems to show Paul, and very likely

Ellen, larking with a "slack wire" music-hall performer, Harry Lamore, in *Fun on the Clothesline* (1897) Although wearing a top hat, Robert seems to be dressed for vigorous performance, as he helps Lamore briefly stay aloft in what looks like a garden setting. Few other surviving films from this period convey as vividly the sense of fun that clearly pervaded early "animated photography."

There was indeed a world waiting to be filmed, and to be turned into the twentieth century's most popular entertainment. But it all lay in the future on that day in North London in 1897.

"A Novel Form of Exhibition or Entertainment, Means for Presenting the Same"

PAUL'S "TIME MACHINE" PATENT
APPLICATION, 1895

My invention consists of a novel form of exhibition whereby the spectators have presented to their view scenes which are supposed to occur in the future or past, while they are given the sensation of voyaging upon a machine through time, and means for presenting these scenes simultaneously and in conjunction with the production of the sensations by the mechanism described below, or its equivalent.

The mechanism I employ consists of a platform, or platforms, each of which contain a suitable number of spectators and which may be enclosed at the sides after the spectators have taken their places, leaving a convenient opening toward which the latter face, and which is directed toward a screen upon which the views are presented.

In order to create the impression of traveling, each platform may be suspended from cranks in shafts above the platform, which may be driven by an engine or other convenient source of power. These cranks may be so placed as to impart to the platform a gentle rocking motion, and may also be employed to cause the platform to travel bodily forward through a short space, when desired, or I may substitute for this portion of the mechanism similar shafts below the platforms, provided with cranks or cams, or worms keyed eccentrically on the shaft, or wheels gearing in racks attached to the underside of the platform or otherwise.

Simultaneously with the forward propulsion of the platform, I may arrange a current of air to be blown over it, either by fans attached to the sides of the platform, and intended to represent to the spectators the means of propulsion, or by a separate blower driven from the engine and arranged to throw a regulated blast over each of the platforms.

After the starting of the mechanism, and a suitable period having elapsed, representing, say, a certain number of centuries, during which the platforms may be in darkness, or in alternations of darkness and dim light, the mechanism may be slowed and a pause made at a given epoch, on which the scene upon the screen will come gradually into view of the spectators, increasing in size and distinctness from a small vista, until the figures, etc., may appear lifelike if desired.

In order to produce a realistic effect, I prefer to use for the projection of the scene upon the screen, a number of powerful lanterns, throwing the respective portions of the picture, which may be composed of,

(1) A hypothetical landscape, containing also the representations of the inanimate objects in the scene.

(2) A slide, or slides, which may be traversed horizontally or vertically and contain representations of objects such as a navigable balloon etc., which is required to traverse the scene.

(3) Slides or films, representing in successive instantaneous photographs, after the manner of the Kinetoscope, the living persons or creatures in their natural motions. The films or slides are prepared with the aid of the Kinetograph or special camera, from made up characters performing on a stage, with or without a suitable background blending with the main landscape.

The mechanism may be similar to that used in the Kinetoscope, but I prefer to arrange the film to travel intermittently instead of continuously and to cut off the light only during the rapid displacement of the film as one picture succeeds another, as by this means less light is wasted than in the case when the light is cut off for the greater portion of the time, as in the ordinary Kinetoscope mechanism.

(4) Changeable coloured, darkened, or perforated slides may be used to produce the effect on the scene of sunlight, darkness, moonlight; rain, etc.

In order to enable the scenes to be gradually enlarged to a definite amount, I may mount these lanterns on suitable carriages or trollies, upon rails provided with stops or marks, so as to approach to or recede from the screen a definite distance, and to enable a dissolving effect to be obtained, the lantern may be fitted with the usual mechanism. In order to increase the realistic effect I may arrange that after a certain number of scenes from a hypothetical future have been presented to the spectators, they may be allowed to step from the platforms, and be conducted through grounds or buildings arranged to represent exactly one of the epochs through which the spectator is supposed to be traveling.

After the last scene is presented, I prefer to arrange that the spectators should be given the sensation of voyaging backwards from the last epoch to the present, or the present epoch may be supposed to have been accidentally passed, and a past scene represented on the machine coming to a standstill, after which the impression of traveling forward again to the present epoch may be given, and the re-arrival notified by the representation on the screen of the place at which the exhibition is held, or of some well-known building which by the movement forward of the lantern can be made to increase gradually in size as if approaching the spectator.

ROBT. W. PAUL
24 October 1895

Flotation Advertisement, 1897

The following appeared in a number of newspapers in London and the provinces on 25 April 1897 in the column headed Public Companies. It was apparently unsuccessful in attracting enough investors, and no record has been found of any further attempt to float a limited company. Since Paul was not obliged to file accounts for a private company, it provides the only financial account of his business up to March 1897, as well as a forecast of how the moving-picture business might develop.

✳

The Public Subscription LIST will OPEN on April 26, and CLOSE on April 27 for London and April 28 for the Country.

PAUL'S AMIMATOGRAPHE,

LIMITED

Capital £60,000,

in 15,000 Ordinary Shares of £1 each, and 45,000 7 per cent Cumulative Preference Shares of £1

DIRECTORS.

John Hollingshead, Esq. (Chairman), Tower Grounds, New Brighton, and 8 Egerton-mansions, Brompton W (formerly of the Gaiety and Alhambra Theatres).

Nevil Maskelyne, Esq. Egyptian-hall, Piccadilly.

Arthur Rayment, Esq, of Messrs Perken, Son, and Rayment, 9 Hatton-garden, E.C.

*Robt. W Paul M.I.E.E, Electrician and Manufacturer of Scientific
Instruments to H.M. Government, 44 Hatton-garden and
114–15 Saffron-Hill, E.C.

*Will join the Board after allotment.

BANKERS.

The National Bank, Limited, 13 Old Broad-street, E.C.

Secretary and offices (pro tem)

George Francis Taylor, 44, Hatton-garden, E.C.

PROSPECTUS.

This company has been formed for the purpose of acquiring the
valuable inventions and patent rights, together with the important
and extensive business, connected with

PAUL'S ANIMATOGRAPHE,

Or Theatrograph,

Which is causing such widespread sensation by the display of
animated photographs in the principal places of amusement, and
to develop the resources of the invention and extend its present
lucrative field of operations in various ways, notably in the
directions of

ANIMATED ADVERTISEMENTS

AND

ANIMATED PORTRAITURE.

The business will be taken over as a going concern, and its
enormously profitable nature is proved by the fact that even during
the first year, i.e. March 2nd 1896 to March 17th, 1897, in spite of the
initial expenses and difficulties of manufacture, which have been
overcome, it has yielded a

NETT PROFIT of £12,838 15s. 4d

On a capital of about £1,000 or 1,200 per cent. There is every reason
to believe that this nett profit, large as it already is, may be greatly
increased by the new developments proposed to be undertaken
by the company, and for which the working capital will provide
adequate means.

The books and accounts relating to the business have been submitted to Messrs Jackson, Pixley, Browning, Husey and Co, Chartered Accountants, who report as follows: —

To the DIRECTORS OF PAUL'S ANIMATOGRAPHE, Limited

Gentlemen, — we have examined the books of Mr R. W. Paul from the 2ⁿᵈ March, 1896 (the date of the first entry in his books relating to Animatographes or Theatrographs), to the 17ᵗʰ March, 1897, and have ascertained that the income therefrom, including sales and receipts for Exhibitions at Theatres, Music Halls &c, amounted to £18,797 3s 4d.

Owing to the fact that, in Mr Paul's Account Books the expenditure of the above business is amalgamated with that of his own business of instrument maker, we have not been able to exactly apportion payments for certain minor purchases, wages etc. We have however ascertained the amount of the principal purchases, and after making ample provision for the purchases and wages referred to, and an amount to cover contingencies, we certify that the net profits for the abovementioned period amounted to £12,838 15s 4d.

The assets to be acquired by the company comprise the patent rights of the Animatographe in Great Britain, Belgium, Austria, Hungary, and Germany, the plant for the manufacture of animated photographs contained in the factory at 36 Leather Lane, a large number of unique negatives (many of which could not be replaced at any cost), a large stock of Animatographes, accessories, and films for animated photographs. The company will also secure the goodwill, together with contracts for exhibitions extending to Christmas, 1897, the benefits of all improvements in progress and to be made hereafter, trade secrets, and the British patent rights of a New Model of the Animatographe, invented by Mr Nevil Maskelyne, to be acquired by the Company in accordance with the Agreement, dated 17ᵗʰ April, 1897, hereinafter mentioned. [...]

Dear Sirs, — In reply to your inquiry, we beg to say that we duly obtained the complete acceptance of Mr Paul's British application (no. 4,688 of 1896), and the Patent was sealed on the 6ᵗʰ inst.

The Austrian Patent was issued on the 14ᵗʰ July, 1896; the Hungarian Patent was issued on 21ˢᵗ October; and the Belguan Patent on 15ᵗʰ May, 1896.

The German application, filed on the 17ᵗʰ April, 1896, was allowed on the 19ᵗʰ February, 1897. Yours truly, D. Young & Co

The Company's business will include the following branches:

1.   THE MANUFACTURE AND SALE OF
     ANIMATOGRAPHES AND ACCESSORIES.

The nett profit on this branch ... amounted to £6,585 8s 6d, notwithstanding the small capital employed, and the initial expense necessarily incurred in bringing out a new invention and establishing a new industry.

The factory and plant in Leather-lane are capable of an output of animated photograph films, of a saleable value of over £200 per week, the net profit on which is over 100 per cent.

An important extension of business will probably result from the supply to amateurs and lanternists of an instrument cheaper and simpler than that required for public entertainments. For use with this, the subjects no longer required for public exhibition may be readily sold.

2.   THE MANUFACTURE AND SALE OF ANIMATED
     PORTRAITS OF INDIVIDUALS.

This great and taking novelty, produced at small cost, preserves for all time and in a portable and convenient form, records of the natural movements of individuals. It is intended to open studios in London and in the principal Provincial Towns for the purpose of taking these portraits of the general public, and to license country photographers to take negatives to be printed in the Company's factory. The profits of this branch are estimated to produce a handsome return.

3.   ANIMATED ADVERTISEMENTS.

One of the most striking results of Mr Paul's inventions, and probably the most profitable, is the production of an animated advertisement, adaptable to any trade and showing processes of manufacture or use of any commodity in life-like operation. Suitable stations for such advertisements are available at moderate cost, and several different advertisements may be shown at each station. The novelty and attractiveness of such advertisements will probably cause them to supersede all other illuminated devices. Several prominent advertisers are negotiating for such advertisements on extremely remunerative terms.

Starting on a basis of only 30 advertisements at £3 each

per weeks for each station, each advertiser taking an average of five positions, this branch alone would produce an income of £23,0000 per annum, which would give, after a most liberal allowance for all expenses and contingencies, an estimated annual profit of about £15,000 per annum.

4.   THE PUBLIC EXHIBITION OF THE ANIMATOGRAPHE.
The Animatographe is one of the greatest attractions the world of amusement has ever had, and it is enjoying phenomenal success, ever increasing as new and exciting subjects are added. Its great reputation is all parts of the world, and the sensation it causes, may be judged form a pamphlet to be had on application, containing Thousands of Laudatory Press notices; and its continued popularity is insured by the possession of a unique set of subjects, photographed in England and abroad, and will be maintained by the constant production of novelties. For instance, a series representing "A Tour in Egypt" has just been photographed, and arrangements have been made for photographing the Diamond [Jubilee] Procession and Festivities.

The nett profits of this branch for 1896–97 amounted to £6,253 6s 10d. The Contracts in hand for Music Hall and Theatrical Exhibitions extend to the end of 1897. The Animatographe, after exhibition for 51 weeks at the Alhambra and 45 weeks at the Canterbury and Paragon Theatres, is now at the "Oxford" Music-hall. The receipts from provincial contracts during the last three months have been more than double those from the previous three months.

A new and extensive field will be opened up by the travelling shows now in preparation, which it is proposed shall contain their own means of locomotion, and have the whole apparatus set up ready for work with electric plant; this will enable side shows to be worked in country towns and fairs and at the forthcoming Jubilee Fetes to be held throughout the country.

Examples of receipts from side-shows—
Olympia, April 4th, 1896, one day £63,160
Edinburgh Festival, week ending January 9th, 1897 .... £132,123

The sale of foreign patents, and the granting of licences abroad may be regarded as a very important source of revenue

Estimate profits of working for the year ending March 31st, 1898, are as follows:

Manufacture and sale of Animatographes and Accessories £7,000

Exhibitions of Animatographes and Travelling Shows 7,000

Profits in Animated Portraits, say 1,000

Profits on Animated Advertisements, first year, say 15,000

.... £30,000

Owing to the rapid progress of the undertaking and the extensive nature of the developments about to be made, Mr Paul is no longer able, unaided, to cope with the enterprise in such a manner as to do justice to the great possibilities of his inventions, and considers it advantageous to offer the concern for public subscription, as far as possible among entertainment proprietors, advertisers, photographers, and others interested in pushing the company's business. Mr Paul undertakes to act as Managing Director for a period of five years.

The purchase price of the business is £25,000 cash and £15,000 fully paid Ordinary Shares, which will not be entitled to rank for any dividend till 7 per cent has been paid on the Preference Shares. After payment of a 7 per cent dividend on the Preference and Ordinary shares, half the surplus profits will be divided pro rata among the Preference Shareholders, and the balance between Ordinary shareholders.

Regarded as an investment, the undertaking bears conclusive evidence of security; whilst with the ample capital to be provided by this issue, the possibilities are enormous, without involving more than ordinary commercial risk. [...]

If no allotment is made, the depsits will be returned without deduction.

Robert Paul Productions 1895–1909

This list aims to record as many as possible of the films that Robert Paul made, commissioned, and "produced." It is based on a number of sources, primarily Paul's early lists and later, more elaborate catalogues. The catalogues were at their most detailed around 1900–1901, but unfortunately not all survive (the British Film Institute holds the largest number). For the period 1894–1900, John Barnes's five-volume *The Beginnings of Cinema in England* provides the most detailed and inclusive listing. Other sources include Dennis Gifford's two-volume *British Film Catalogue* and the ever-expanding list under "R. W. Paul" on IMDb; the latter, unfortunately, lacks sources or corroboration, and overconfidently specifies "directors."

Films currently known to survive are asterisked, with the mark in parentheses where identification is uncertain. Dates usually refer to the first mention of the film, whether in an ad, a catalogue, or occasionally a press notice. Original/advertised lengths are given in feet, although surviving versions are often shorter, as with most early films. For pre-1910 films running at 16–18 frames per second, 100 feet would last about 1.5 minutes, while the 40-foot length common in 1896–1898 would have run for about 40 seconds.

1895

Title/description	Date	Length
Cricketer Jumping over Garden Gate	95 Feb	
Oxford and Cambridge Boat Race	Mar	
Incident outside Clovelly Cottage	Mar	
*Arrest of a Pickpocket	Apr	
The Derby	May	
Comic Shoe Black/Shoeblack at Work in a London Street	May–Jun	
Boxing Kangaroo	May–Jun	
Performing Bears	May–Jun	
Boxing Match	?	

(continued)

1895 *(continued)*

Title/description	Date	Length
Carpenter's Shop/Beer Time	May–Jun	
Dancing Girls	?	
*Rough Sea at Dover	Jun	
Tom Merry, Lightning Cartoonist	?	

Following Barnes, these 1895 films—made for the Kinetoscopes Paul was manufacturing, with the camera Paul and Birt Acres had developed—are designated "Paul-Acres." A crime drama known as *Footpads* was included on the British Film Institute DVD *R. W. Paul: Collected Films*, assigned to 1895–1896. There is, however, no corroborating evidence for the attribution, and the film's scenography differs from that of any known Paul production. See Ian Christie, "Issues of Provenance and Attribution for the Canon: Bookending Robert Paul," *Early Cinema Review*, proceedings of the 2018 Domitor conference (Bloomington: Indiana University Press, 2020).

1896

Title	Date	Length
Smith[s] and Machinery at Work/Nelson Repairing Dock	Feb	40
Dancing Girls (Alhambra girls; colored by Doubell)	Mar	
Contortionist	Mar	
Trilby Burlesque (Alhambra Girls)	Apr	
A Crowd in the City	Apr	
On Westminster Bridge	Apr	
*The Soldier's Courtship/very funny	May	
Rough Sea at Ramsgate	May	
Rough Sea at Ramsgate, no. 2	May	
Arrival of the Paris Express at Calais	May	
Costermonger	May	
The Derby (2 parts)/exciting victory of Prince's horse	Jun	
On the Calais Steamboat [Yvette Guilbert?]	Jun	
Hampstead Heath on Bank Holiday/lively scene	Jun	
*David Devant, the Conjuror/The Mysterious Rabbit	Jun	
Devant produces eggs from all parts/The Egg-Laying Man	Jun	
Devant's Exhibition of Paper Folding	Jun	
Devant's Hand Shadows (2 parts)	Jun	
*Chirgwin, the White-eyed Kaffir, in his humorous business	Jun	
Chirgwin Plays a Scotch Reel on tobacco pipes	Jun	
Landing at Low Tide	Jun	
The Tea Party	Jun	
*Traffic on Blackfriars Bridge; beautiful detail	Jun	
Mr Maskelyne Spinning Plates	Jul	
*Up the River/exciting rescue of child fallen overboard	Jul	
Henley Regatta (2 parts)	Jul	
A Comic Costume Race/Music Hall Sports	Jul 14	
*The Twins' Tea Party/two infants at tea; very amusing	Aug	
*Two a.m.; or, The Husband's Return	Aug	
Arrest of a Bookmaker	Aug	
Morris Cronin, marvellous American Club Manipulator	Sep	
Cronin with Three Clubs; a wonderful feat	Sep	
The Sisters Hengler Speciality Dancers	Sep	
Children in the Nursery	Oct	
You Dirty Boy ("Famous statue brought to life")	Oct	

1896 (continued)

Title/description	Date	Length
Princess Maud's Wedding (2 parts)	Aug 8	
Juvenile Plate Race/Music Hall Sp	Aug 8	
Children at Play no.1/children and dogs at play	Aug	
Children at Play no.2 [parents watching]	Aug	
[On Brighton Beach] humorous scene, landing from boat	Aug 8	
The Gordon Highlanders (3 parts, Maryhill Barracks)	Aug 22	
Fred Holder and Family [Canterbury manager]	Aug 22	
SS *Columba*/Leaving Rothesay Pier Aug 15	Sep 5	
Passengers Disembarking SS *Columba*	Sep 5	
Gardener Watering Plants	Sep 21	
Buchanan Street, Glasgow	Aug?	
Argyll Street, Glasgow	Sep	
Omnibuses & pedestrians passing St Thomas' Hospital	Oct 3	
A Tour in Spain and Portugal (series)	Oct 10	
Lisbon, the Fish Market		
Madrid, Puerto del Sol		
The Bathers/lively scene of diving and bathing nr Lisbon		
Cadiz, the Square/Plaza del Cathedrale		
Seville, Bull Fight		
*Andalusian Dance/two performers		
Seville, Leaving Church of San Salvador/After Mass		
Portuguese Game of Pau/two men actively engaged		
The Fado		
Lisbon, Praca da Municipio		
Lisbon, Praca Dom Pedro		
Portuguese Railway Train		
*Sea Cave Near Lisbon/artistic view of/waves violently rushing in		
Seville, The Triana		
Paper Makers	Nov 23	
Fish Wives	Nov 23	
Hyde Park Bicycle Scene	Nov 23	
Turn-out of a Fire Brigade/most exciting	Nov	
Football Match (Newcastle on Tyne)	Nov	
Petticoat Lane/Sunday morning in Whitechapel	Dec 12	
Feeding Pelicans at the Zoo	Dec 19	
The Lord Mayor's Show (2 parts, Nov 9)	Dec	
Brighton Pier	?	
The Doomed Chimney Stack/underpinning	?	
Birmingham	?	
Edinburgh	?	
Sunderland	?	
Houses of Parliament	?	
*Royal Train	?	
Terrible Railway Accident ?	?	
Scene on the SS *Victoria*, from Dover to Calais	?	

1897

Title/description	Date	Length
Carting Snow/shovelling snow in Leicester	Feb 13	
Egypt (series)		
Arab knife grinder at work	Apr 24	
Arab Exercising his Monkey		
*Women Fetching Water from the Nile		
Caravan Arriving at Pyramids		
Cairo Scene: Selling Water from a goat skin		
Loading Camels in a Cairo Street		
Arabs at Work Breaking Old Iron with primitive machinery		
Egyptian Bullock Pump; drawing water		
*Fisherman and Boat at Port Said		
Egyptian Scene: Primitive Saw Mill		
Procession of the Holy Carpet from Cairo to Mecca		
Dervishes Dancing		
An Egyptian Cigarette Maker at Work		
Queen Victoria's Diamond Jubilee Procession (series)	Jun 26	
*Head of Procession including Bluejackets		40
*Head of Colonial Procession, Canadians		40
*Royal Carriages Passing Westminster		40
Colonial Troops Passing Westminster		40
Continuation of Colonial Troops (joins no. 101)		40
*Cape Mounted Rifles Passing St Paul's		40
*Dragoons Passing St Paul's		40
*Royal Carriages Arriving at St Paul's		40
Life Guards and Escort Arriving at St Paul's		40
*Queen's Carriage and Indian Escort Arriving		40
*Life Guards and Princes North of St Paul's		40
Queen and Escort in Churchyard, showing Queen's face		40
The Derby/a fine picture	Jul	
Mountainous/scene from play *The Geisha*	Sep	40
Marquis/scene from *The Geisha*	Sep	40
The Miller and the Sweep		
*Cupid at the Washtub		
The Gamblers		
The Young Rivals		
The Rival Bill-Stickers		
A Lively Dispute		
Theft/two tramps steal goose, pursued by farmer's wife		
Spree/corporal catches soldiers in public house and joins them for a drink		
Goat/soldiers dress goat in their uniforms, and it butts them		
Quarrelsome Neighbours/neighbours fight over fence with soot and whitewash		
Robbery/a wayfarer compelled to disrobe partially		60
Jealousy/dramatic scene in gardens—jealous husband shot		80
Village Blacksmith/Mel B. Spurr recites Longfellow poem		60
*Fun on the Clothesline		60
(*)The Vanishing Lady/with Charles Bertram		40
Hail Britannia!/conjuror produces flags that change into Britannia		60

1897 (continued)

Title/description	Date	Length
City and Suburban Handicap/Apr 21	Oct	40
Prince of Wales Reviewing Yeomanry/Chelt/May	Oct	120
Fire Brigade Review at Windsor/Jun 28	Oct	40
Union Jack fluttering in the breeze	Oct	40
Chimney/Jack Smith underpins & overthrows stack	Oct	80
Traffic on Tower Bridge	Oct	40
Fountains/Trafalgar Square	Oct	40
Sweden (series)		
Passengers arriving at Stockholm	Oct	40
Laplander Feeding his Reindeer		
Laplanders Arriving at their Village; welcome home		
Children at Play in the King's Garden		
Derricks at Work Discharging Coal from steamer		
Electric Trolley Car coming through pine forest		
King's Guard Marching at Stockholm		
Steamboat Aeolus Leaving Stockholm		
Swedish National Dance at Skansea		
Swedish National Dance; eight Performers		
Swedish National Dance; four performers		
Bank Holiday Picture at Hampstead/Aug 1	Oct	40
Ferries Arriving at Landing Stage/B'head,L'pool	Oct	60
Douglas, Isle of Man (series)	Oct	60
Lively Picture at Douglas: Diving		80
Landing Visitors from Small Boats		40
Laxey Electric Railway		
Scene on Douglas Beach		
SS *Empress Queen*		
Clock Tower		
The Parade		
Douglas Ferry		
Diving and Bathing in the Sea/Brighton	Oct	40
Donkey Riding on the Beach/Brighton	Oct	40
Rottingdean Electric Railway/Sea-going car	Oct	80
Meet of Staghounds at Aylesbury/Lord Rothschild's pack	Oct	40

1898

*Clog Dancing Contest for Championship of England	Feb 19	120
Express Train on Forth Bridge	Mar 19	40
Launch of the Braemar Castle/Launching a Liner/Feb 28	Mar 19	40
Hunting	Mar 19	40
Greenwich/panorama pier and hospital	Mar 19	40
Unloading Cargo at London Docks	Mar 19	40
Panorama of London Streets	Mar 19	40
Fregoli The Protean Artiste/impersonations of musicians	Mar	400
The London Express/GNR train passing Wood Green	Mar	40
Gladstone's Funeral/Fun. of late W E Gladstone May 28	Jun 4	150
The Derby 1898/Jeddah wins	Jun 4	80
Trooping the Colour/May 24	Jun 4	60

(continued)

1898 (continued)

Title/description	Date	Length
Quorn Hunt/Leics meet at Kirby Gate	Jun 11	80
Shipping on the Thames/pan from tug	Jun 18	40
Steamers/pan from tug	Jun 18	40
Cory/pan of floating derrick	Jun 18	40
*Launch of HMS *Albion*/battleship sinks 21 Jun	Jun 25	40
The Blackwall Disaster	Jun 25	40
SS *Carisbrooke Castle* (series)	Jun 25	40
Naval Games		
Pillow Fight, or Tournament		
Tether Ball, or Do-Do		
A Switchback Railway	Jul	40
Fire at Alexandra Palace	Jul	60
Fire Drill/soldiers drill at Bedford Barracks	Jul	60
A Rescue from Drowning	Jul	80
Rescued/spectators cheer rescue	Jul	40
*Comic Costume Race/Music Hall Sports 5 Jul	Jul	40
Burlesque Football/walking dog race	Jul	40
Fancy Dress Cycling	Jul	40
Obstacle Race	Jul	40
Constantinople (series)		
Embarking	Aug cat	40
Scene of Armenian Massacres	Aug	40
A Street in Stamboul	Aug	40
Street Scenes (series)		
Edinburgh	Aug	40
Sheffield		
Brighton		
Burnley		
Hull		
Liverpool		
Cardiff		
London (series)		
Piccadilly Circus; Blackfriars Bridge	Aug	40
Westminster Street Scene		
Outside the Paragon		
Outside the Oxford		
Panorama of Holborn		
Leicester (series)		
Clock Tower	Aug	
London Road		
Pantomime Rehearsal		
Glasgow (series)		
Buchanan Street	Aug	
Sauchiehall Street		
Rothesay Pier		
Argyll Street		
A Study in Facial Expression/man laughs at risqué story	Aug	60
The Stockbroker/ruined broker tries to shoot himself	Aug	60
Birdsnesting/boy falls from tree on farmer	Aug	60
When the Cat's Away/boy playing tricks in class	Aug	60

1898 (continued)

Title/description	Date	Length
Photography/pests interfere with photographer	Aug	60
The Sailor's Departure/sailor taking leave of wife	Aug	60
The Sailor's Return/returns and sees their baby	Aug	60
The Monks/one smoking cigar burns sleeping friend	Aug	60
Jovial monks in the Refectory/novice gets drunk	Aug	80
Mr Bumble the Beadle/courting scene in *Oliver Twist*	Aug	60
Old Time Scene in Village Stocks	Aug	80
The Breadwinner/poor mother dresses child as fairy	Aug	80
The Fairy/"similar to The Breadwinner but shorter"	Aug	60
*Tommy Atkins in the Park/open air version	Aug	80
*A Favourite Nursery Scene/girl hides under bed	Aug	80
Rescue from Drowning/people rescued by rowboat	Aug	80
The Lodger/attacked by fleas	Aug	80
In the Name of the Queen/deserter hidden by mother	Aug	80
*Come Along, Do! /2 sc; country couple at gallery	Aug	80
Our New General Servant/4 scenes; first intertitles	Aug	320
Arrival of the Sirdar at Dover	Oct 29	
Grenadier Guards Returning from the Soudan	Oct 29	60
*Reception of the Sirdar at Guildhall 4 Nov	Nov 12	80
Ashore and Afloat (series?)		
Passengers Embarking for South Africa	Nov	
Santa Claus and the Children/presents down chimney	Nov	60
RMS Hawarden Castle/leaving for Cape	Dec 24	
Slow Trains and Fast	Dec	40
Hay/building haystack	Dec	40
Highland Fling/Scots soldiers dance in camp	Dec	40
Parliament Street, London	Dec	60

1899

A Storm in Dover Harbour	Apr 1	80
Visit of the Queen to South Kensington/V&A foundation	May 27	80
The 1899 Derby	Jun 10	40
Trooping the Colour at Horse Guards	Jun 10	120
Photographic Convention, Gloucester	Jul 21	150
Bertie's Bike; or, The Merry Madcaps	Sep	50
The Country Waiter; or, The Tale of a Crushed Hat	Sep?	70
Two Tipsy Pals and the Tailor's Dummy	Sep?	60
A Gretna Green Wedding	Sep?	60
The Bricklayer and his Mate; or, A Jealous Man's Crime	Sep?	50
Thrilling Fight on a Scaffold/while building studio	Sep?	100
The Miser's Doom/ghost frightens miser	Sep?	45
On a Runaway Motor Car through Piccadilly Circus	Sep	80
*Upside Down; or, The Human Flies	Sep	80
An Unexpected Visit; or, Our Flat/faking a chair	Sep	70
Caught Flirting/by girl's father, who sets dog on suitors	Sep	100
Shooting a Boer Spy	Nov	60
The Bombardment of Mafeking	Nov	60
The Battle of Glencoe	Nov	80

(continued)

1899 (continued)

Title/description	Date	Length
*Surprising a Picket/Attack on a Picquet	Nov	40
*A Camp Smithy	Nov 25	60
The Kaiser at Portsmouth/also Nelson's Flagship	Nov 25	80
British Capturing a Maxim Gun	Dec	60
Nurses Attending the Wounded/On the Battlefield	Dec	60
The Victory	Dec	60
Wrecking an Armoured Train	Dec	100
Fun on a Transport	Dec 9	
Scots Guards Embarking/on SS *Nubia* at Southampton	Dec 9	60
The Siege Train	Dec 9	

1900

	Date	Length
Snowballing Oom Paul/children make Kruger snowman	Jan	40
Modder River Drift/Ambulance crossing Modder	Jan 6	80
Bridging the Modder River/replacing wooden bridge	Jan 6	60
Mule Wagons Crossing the Modder	Jan 6	40
Naval 47 Gun drawn by 40 Bullocks	Jan 13	60
Cavalry Watering their Horses in the Modder	Jan 13	40
City Imperial Volunteers Leaving Wellington Barracks	Jan 13	100
Lord Mayor at Southampton	Jan 13	
(*)Royal Engineers' Balloon	Jan	60
Embarkation of the CIVs for South Africa	Jan 13	50
Dragging up the Guns	22 Jan	80+100
Transporting Provisions to the Front	Jan 24	50
Crossing the Vaal River/Lord Roberts and Scots Guards	Jan	55
Artillery Crossing the River Vaal	Jan	58
Ambulance Train	Jan	60
Telegraphing Casualties/field clerks & Wheatstone auto	Feb 3	50
Hurrah for the Queen/Xmas Day 1999	Feb 3	40
Naval Gun/"Joe Chamberlain" answers "Long Tom"	Feb 3	60
Fording Modder River	Feb 3	50
Magersfontein/ammo wagon & gun Jan 22	Feb 24	100
Boer Shell-Proof Pits	Feb/Apr 6	
Plucked from the Burning (2 scenes)	Mar 31	100
The Hair-Breadth Escape of Jack Sheppard	Mar 10	100
Battle of Poplar Grove	Mar 7/Apr 14	
Diving for Treasure	Mar 10	45/120
*A Railway Collision	Mar 10	40
The Brutal Burglar	Mar 10	
*Scots Guards Triumphal Entry into Blomfontein	Mar 13 Apr 14	120+80
A Morning at Bow Street/"Four lively cases"	Mar 31	
The Worried German	Mar ?	90
High Life Below Stairs	Mar ?	80
The Queen's Visit to London/at Woolwich?	Mar 10/14.4	40 + 60
*Cronjé's Surrender	Mar 31	60
Review of Berks Volunteers by Queen, Feb 28	Apr 7	50 + 60
HMS *Powerful* arriving at Portsmouth	Apr 14	80
Boat Race 1900	Apr 14	
Football Match	Apr 14	

1900 *(continued)*

Title/description	Date	Length
Sir George White's Arrival	Apr 21	50
Great Review by the Queen in Phoenix Park	Apr 28/4	120
[As above, showing only Queen with escort]	Apr 4	
Landing of the Naval Brigade Apr 11	Apr 28	40
Return of the Naval Brigade which Saved Ladysmith	Apr 28	40
To the Paris Exposition by Newhaven-Dieppe	Apr	100
The Naval Brigade in London/from Victoria 7.5	May 19	
Kruger's Dream of Empire	May	65
A Naughty Story/facial	May	60
Queen's Birthday Celebrations (4 sc)	May 26	120
War Carnival	May 26	120
Artillery Crossing a River	June	58
The 1900 Derby	Jun 2	40
Prince of Wales interviewing Soldiers' Wives	Jun 30	40
Briton v. Boer	Jul 7	100
Scouting the Kopjes	Jul 21	
His Mother's Portrait; or, The Soldier's Vision	Jul 21	100
Punished/fidelity comedy	Jul 28	40
The Last Days of Pompeii	Jul 28	80/65
Khedive at Guildhall	Jul 7	50
Crossing the Vaal	Jul 28	55
The Royal Engineers' Balloon	Jul 28	60
Gun/sailors drilling at Chatham	Jul 18	80
Hindoo Jugglers	Jul 28	70
Naval Gun Crossing the Vaal River	Aug 4	50
Mule Transport in Ravine nr Pretoria	Aug 4	50
Cowes Regatta (series)		
Pan of Cowes Front Jul	Aug 4	80
Race for Royal Yacht Sq Cup		40
Meteor Striking HM The King's Yacht		60
Racing at Cowes		80
Queen's Yacht, *Victoria and Albert*		50
Chinese Magic/aka Yellow Peril	Sep 15	100
Army Life; or, How Soldiers Are Made (series)	Sep 15	
Series I		
Joining the Army		160
Series II		
Life at the Regimental Depot		
Drilling the Awkward Squad		120
Work and Play at the Depot		80
Series III		
Camp Life at Aldershot		
Firing at the Ranges		60
Soldiers' Bathing Party		60
Dinner in Camp		80
A Lark in Camp		60
Series IV		
Army Gymnastics at Aldershot		
Pyramids		100
Vaulting Horses		80

(continued)

1900 (continued)

Title/description	Date	Length
Series V		
Leaving the Army/Commissionaire Corps		60
Series VI		
"Back to the Army Again"/reserves		50
Series VII		
Training of Cavalry at Canterbury		
Recruit's First Ride		80
Sword Exercise and Lance Drill		100
Bare-back Riding and Charge of the Lancers		80
Musical Ride by the 2nd Life Guards		40
Series VIII		
Royal Army Medical Corps		100
Series IX		
Royal Horse Artillery at Woolwich		60
Watering Horses and Hooking-In the Gun		100
Series X		
Army Service Corps		
Soldier Bakers and Butchers		80
Dismounting a Service Wagon		80
Series XI		
Royal Garrison Artillery		
Firing a 9-in Muzzle Loader		100
Firing a 6-in Disappearing Gun		100
Battery of Quick Firing Guns in Action		80
Series XII		
*The Training of Infantry/Mounted Infantry		40
Digging Trenches		40
Soldier Cyclists in Action		70
Maxim Gun Drill		60
Series XIII		
Work of the Royal Engineers		
Escalading and Capturing a Fort		120
Building a Pontoon Bridge		100
Defending a Redoubt		60
Exploding a Land Mine		30
Exploding a Submarine Mine		100
Constructing a Trestle Bridge		100
Britain's Welcome to Her Sons (5 scenes)	Sep 15	100
A Wet Day at the Seaside/making "seaside" in the bathroom	Sep 15	50
The Drenched Lover/suitor chased into pond	Sep	70
Return of the CIVs/Parade through London	Oct 29	120

1901

	Date	Length
Funeral of Her Late Majesty Queen Victoria (series)		
dep Cowes	Feb 9	120
The London Procession	Feb 9	120
" " "taken by a special lens"	Feb 9	150
Oxford and Cambridge Boat Race 1901	Mar	80
Departure of Duke and Duchess of York	Apr cat	50

1901 (continued)

Title/description	Date	Length
R.M.S. Ophir Leaving Portsmouth/royal yacht	Apr	80
Football Final/Spurs vs Sheffield United, Crystal Palace	Apr	120
Plaiting the Maypole/children's trad dance	May	50
Scenes during Regatta Week (series)		
Cowes Panorama	May	80
Yacht in Cowes Roads		89
Race for Royal Yacht Squadron Cup		80
The King's Yacht Britannia		80
Racing at Cowes		80
Launch of Shamrock II		50
Sir Thomas Lipton		80
Wreck of the Shamrock II		40
Woodford Cyclists Meet/fancy dress	May	50
Kempton Park Races	May	80
The Derby 1901	Jun	100
Five Years' Derbies (compilation)	Jun	100
Acrobatic Performance—Sells and Young	Aug?	150
*The Deonzo Brothers	Aug?	120
The Captain's Birthday	Aug?	100
*An Over-Incubated Baby	Aug?	80
Garters versus Braces; or, Algy in a Fix	Aug?	90
The Artist and the Flower Girl	Aug?	80
The Automatic Machine; or, Oh what a Surprise	Aug?	100
Handy Andy the Clumsy Servant	Aug?	100
*His Brave Defender	Aug?	100
An Interrupted Rehearsal; or, Murder Will Out	Aug?	100
The Muddled Bill-Poster	Aug?	20
William Tell	Aug?	100
*The Devil in the Studio	Aug?	100
*The Haunted Curiosity Shop	Aug?	140
*Undressing Extraordinary; or, The Troubles of a Tired Traveller	Aug?	200
*The Cheese Mites; or, Lilliputians in a London Restaurant	Aug	70
The Drunkard's Conversion/Horrors of Drink	Aug	80
The Tramp at the Spinsters' Picnic	Aug	70
'Arry on the Steamboat/with song	Sep	180
Ora Pro Nobis; or, The Poor Orphan's Last Prayer/song	Sep	100
*The Waif and the Wizard; or, The Home Made Happy	Sep	90
Britain's Tribute to Her Sons/with song	Sep	150
*The Countryman and the Cinematograph	Oct	50
Punch and Judy	Oct	
*Artistic Creation	Oct	85
The Famous Illusion of De Kolta	Oct	120
Hair Soup; or, A Disgruntled Diner	Oct	60
The Tramp and the Turpentine Bottle; or, Greediness Punished	Oct	70
The Gambler's Fate; or, The Road to Ruin	Nov	200
Mr Pickwick's Christmas at Wardle's	Nov	140
*The Magic Sword; or, A Mediaeval Mystery	Nov	180
*Scrooge; or, Marley's Ghost/13 sc	Nov	620
Two Generations of Britain's Future Kings/Yorks	Nov 9	

(continued)

1901 (continued)

Title/description	Date	Length
Departure of the Royal Yacht	Nov 9	120
London's Reception to Royal Travellers/Marlborough House	Nov 9	80
Panorama of the Glasgow Exhibition	?	50
Opening of Glasgow Exhibition, Duke & Duchess of Fife	?	70
Dolly's Toys/animation by Cooper?	Dec	80

1902

State Opening of Parliament	Jan	80
The Cockfight/coster & gent bet	Feb	50
Race for the Grand National 1902	Mar 22	100
Oxford and Cambridge Boat Race	Mar 22	75
Their Majesties at Chelsea Hospital	Mar cat	40
King's Birthday and Trooping the Colour	May 31	120
Presentation of Colours to Irish Guards by King	May 31	
Firing a Salute on King's Birthday	May	40
A Convict's Daring/4 scenes; reissue '07	Jun	260
Love's Ardour Suddenly Cooled	Jun	53
Little Willie's Coronation Celebrations	Jun	75
Facial Expressions	Jun	80
The Hotel Mystery/trick	Jun	85
King and Queen's Arrival in Fleet Street Jun 8	Jun 14	50
Arrival at St Paul's Jun 8	Jun 14	100
Scene at a Duck Farm	Jun 21	50
Canoeing on the Charles River	Jun 21	100
Educated Chimpanzee	Jun 21	120
The Troublesome Collar/facial	Jul	90
*His Only Pair	Jul	75
The Swells	Jul	50
The Enchanted Cup/7 sc	Jul	300
A B-Oysterous Dispute	Jul	50
Father Thames Temperance Cure/trick	Nov	85
Soap versus Blacking/reversing comedy	Nov	110
*The Extraordinary Waiter/Mysterious Heads	Dec	108

1903

Coronation Durbar at Delhi/Jan 31	Jan	110
*The Delhi Durbar/elephants	Jan 24	240
Joe Barrett v. Louis Anastasi	Jan 31	100
Clog Dancing Contest/Colette wins	Jan 31	175
Sport at the Delhi Durbar/Chitral vs Gulghit	Feb 7	50
Visit of Duke & Duchess of Connaught to Khyber	Feb 7	50
Skiing at Quebec	Feb 14	50
Panoramic Life at Rubie Canyon	Feb 14	100
Opening of Parliament/king at Westminster	Feb 21	120
Return Home of Joseph Chamberlain/Mar 14	Mar 21	100
Launch of Sir Thomas Lipton's Yacht *Shamrock III*	Mar 21	
Sports in India/Indian Animal Tourn; bull, ram	Mar 21	100
Return of Duke and Duchess of Connaught	Mar 28	80

1903 (continued)

Title/description	Date	Length
Race for the Grand National/ph RWP ?	Mar 28	80
Nadji the Hindoo Marvel	Feb	120
Weary Willie's Wiles	Feb	200
The Old Love and the New	Feb	100
*Pocket Boxers	Mar	80
The Dice Player's Last Throw	Mar	160
The Washerwoman and the Sweep	Mar	80
The Great International Football Match	Apr 11	200
Shamrock I and III/trial at Portsmouth	Apr 18	
Indian Tour of Prince & Princess of Wales	Apr (cat)	165
Hunting Big Game in India/elephants	Apr (cat)	70
Lion Hunting in India	Apr (cat)	90
Trout Playing/fishing	Apr (cat)	45
Rabbit Hunting with Beagles	Apr (cat)	85
Race for the Derby 1903/Rock Sand wins	May 30	80
The Beasts of the Earth/Leopards, tigers, lions	May 30	150
The Moat Farm Murder/digging for clues	May (cat)	80
Railway Panorama/Derby to Buxton	Jun 13	200
Inter-Polytechnic Champ Sports/Padd rec	Jun 13	120
Trooping of the Colour by the King	Jun 27	
Bloodhounds Tracking a Convict/3 sc inc rep	Jun	340
The Great Motor Car Derby/Gordon Bennett	Jul 11	150
President Loubet's Visit/at Victoria	Jul 11	170
Norway Revisited with the Animatograph (series)	Jul 18	
The Midnight Sun at Scaro		60
Laplanders at Home		150
Greatest Glacier of Europe		120
Beauty Spots of the North		
Snowclad Mountain Tops		95
Log Rolling on the River Nid		110
Panorama of Hammerfest		80
Celebrated Laatifross Waterfalls		120
Panorama of Bergen		120
Railway Ride from Stalheim to Bergen		
Seven Sisters Waterfalls		
Geiranger and Hardanger Falls		55
Railway Ride from Vossevangen to Bergen		
Panorama of the Town of Molde		60
Laplanders with a Herd of Reindeer		135
Magnificent Falls of Tvindevoss/Norwegian Waterfall		80
The Svartisen Glacier		85
Raftsund Mountains		90
Seagulls	Jul (cat)	80
Sports on a Pleasure Yacht	Jul (cat)	100
Father Neptune/crossing Equator ceremony	Jul (cat)	90
Panoramic Railways Ride/Gt Northern system	Aug 22	3 parts
Swimming Contest/Surrey Docks	Aug 22	
*A Chess Dispute	Aug	80
High Diving at Highgate/with reversing	Aug	60

(continued)

1903 (continued)

Title/description	Date	Length
Yorkshire Fed Convoyeurs/rel 800 pigeons Didcot	Sep 12	50
Great Dock Fire at Millwall	Sep 12	130
America's Cup Races 1903 (series)		
Manoeuvring	Sep 12	120
Yachts Racing towards Camera		70
Railway Life/scenes on Gt Northern	Oct 24	200
Panoramic View from Scottish Express/from train	Oct 24	280
HMS Victory in Collision with the Neptune/Ports	Oct 31	80
The Adventurous Voyage of the Arctic	Nov	600
The Kiddies' Cakewalk	Nov	62
*An Extraordinary Cab Accident	Nov	50
Diavolo's Dilemma/Lilo, Lotto, Otto perf	Nov	100
Marionette Performance/skeleton, chair, harlequinade	Nov	215
Nigger Courtship	Nov	80
The Fine Fisherman/minnow and pike	Nov	60
A Good Catch/stealing salmon	Nov	60
Hauling a Heavy Catch of Fish	Dec 5	115
Tobogganing in Switzerland	Dec (cat)	60
Pushball/teams push large ball	Dec (cat)	60
Life Saving Exercising/demo by LS society	Dec (cat)	120
Pie Eating Contest/boys compete	Dec (cat)	80
Novel Airship/Wm Beedle flies at Alexandra Palace	Dec (cat)	50

1904

Political Favourites/politician cartoons	Jan	150
The Adventures of a Window Cleaner/6 scenes	Feb	280
An Artful Young Truant	Feb	50
An Affair of Outposts/Russo-Jap war scenes	Feb	300
Submarine Mine/Naval and Submarine Manoeuvres	Mar 19	150
An Explosion	Mar 19	50
Street Hawkers/London's oldest street sellers	Mar 19	120
Funeral of Duke of Cambridge/West Abbey; Kensal Green	Mar 26	80; 80
Naval Manœuvres/HMS Trafalgar, Japanese ship Niobe	Mar (cat)	100
Japanese Street Scenes	Mar (cat)	76
Jap versus Russian/symbolic wrestling	Mar	100
All for the Love of a Geisha/7 scenes	Mar	540
A Russian Surprise/Chinese waiter serves bomb	Apr	96
That Terrible Sneeze/old man blows wife's skirt up	Apr	75
Capture and Execution as Spies of Two Jap Officers	Apr	365
The Music Hall Manager's Dilemma/trick	May	260
The Chappie at the Well	May	120
The Talking Head	May	80
The Haunted Scene Painter/trick	May	180
Modern Stage Dances/w Margery Skelly	May	75 + 60
Clever Dances	May	150
*Drat That Boy!	May	90
Funny Faces/facial	May	50
*A Miner's Daily Life/A Collier's Life	Jun 18	320
All the Fun of the Fair	Jun	190

1904 (*continued*)

Title/description	Date	Length
*Buy Your Own Cherries/William West song	Jun	300
The Student, the Soot and the Smoke	Jul	160
That Terrible Barber's Boy	Jul	105
The Ploughboy's Dream/dreams he is a pirate	Jul	136
Looking for Trouble	Jul	104
The Spiteful Umbrella Maker	Jul	106
The New Turbine Torpedo Boat/steaming at 30 knots	Aug 20	90
How Our Coal is Secured/Down a Coal Mine	Aug 20	315
*Mr Pecksniff Fetches the Doctor/Oh What a Surprise	Aug	145
The Sculptor's Jealous Model	Aug	183
Love Laughs at Locksmiths	Aug	150
Why Marriage is a Failure	Aug	120
From London to Penzance (series)		
Pan Windsor & Slough	Sep 17	65
Bath and the Great Roman Baths		100
Racehorse Life on the Berkshire Downs		160
Cheddar with its Perpendicular Cliffs		90
Panorama of the River Fal		90
Seagulls at Newlyn Harbour		55
View of the Cornish Coast		135
Clifton Suspension Bridge		45
Railway Ride from Reading to Bristol		120
The River Dart		65
Dawlish and Teignmouth		100
Scilly Isles		45
Land's End. The Rugged Cornish Coast		135
Unique Railway Rides		200
The Life of a RaceHorse/from Berks Downs to racecourse	Oct 22	160
Russian Outrage in the North Sea (reissues)		
Trawling (series)	Oct 29	60
Mending the Nets		95
Hauling in a Big Catch		115
Washing and Sorting Fish on Board		70
Panoramic View of the Humber		95

1905

The Armless Wonder/Herr Unthank uses feet for everything	Apr 15	190
Papa Helps the Painters	Apr	170
Auntie's First Attempt at Cycling	Apr	200
*A Victim of Misfortune/The Unfortunate Policeman	Apr	250
The Fatal Necklace	Apr	320
A Race for Bed	Apr	200
Elephants at Rangoon	Mar 13	125
Trouble Below Stairs	May	120
The King of Clubs/trick	May	175
The Tramp and the Typewriter	Jun	150
Alyesbury Ducks/feeding and swimming	Jul 1	130
Launch of Japan's New Warship/Jul 4	Jul 8	200

(*continued*)

1905 *(continued)*

Title/description	Date	Length
Ascent of the Barton-Lawson Airship/Alexandra Palace 22.7	Jul 29	110
Living beyond Your Means	Jul	245
The Great Channel Swim/Jabez Wolff	Aug 26	120
While the Household Sleeps	Aug	315
He Learned Ju-Jitsu—So Did the Wife	Aug	335
The Adventures of a £100 Banknote	Aug	100
The Pierrot and the Devil's Dice	Aug	175
Royal Review of Scottish Volunteers/Edinburgh Sep 18	Sep 23	150/250
Goaded to Anarchy/Russian anarchist	Sep	480
The Freak Barber/black and white men swap heads	?	168
The Dancer's Dream	Sep	180
When the Wife's Away	Sep	250
Madrali v. Jenkins/wrestling at Lyceum Oct 2	Oct 7	145
The Life of a Railway Engine/GNR Works Doncaster	Oct 14	200
Short-Sighted Sammy	Oct	
The Visions of an Opium Smoker	Oct	262
The Misguided Bobby	Oct	120
Elephants at Work/Rangoon River	Dec (cat)	125
Club Swinging/Tom Burrows at St Georges Hall	Dec (cat)	175
Ball Punching/Ernest Plummer demonstrates	Dec (cat)	130
A Christmas Card; or, The Story of Three Homes	Dec	215
A Shave by Instalments on the Uneasy System	Dec	267

1906

An Indian Animal Tournament/Prince watches bull v ram	Jan 13	100
Prince, Princess of Wales at Gwalior/visiting Maharajah	Jan 20	160
Arrival of Prince of Wales at Calcutta/inspect guard	Jan 27	100
England v. Ireland: match at Leicester	Feb 10	
The Doctored Beer/dir. by J. H. Martin?	Mar	260
Mistaken Identity/Martin?	Mar	105
He Cannot Get a Word in Edgeways/Martin?	Mar	170
The Old Lie and the New/Martin?	Mar	100
The Fakir and the Footpads/Martin?	Mar	214
*A Lively Quarter Day	Mar	332
The Madman's Fate/w Leah Marlborough 11 sc	May	576
Seaside Lodgings	May	500
The Race for the Derby/Spearmint's Derby May 30	Jun 2	115
Coronation of King Haakon VII and Queen Maud	Jun (cat)	200
Norwegian Cabmen at Gudvangen	Jun (cat)	80
Jabez Woolffe the Channel Swimmer	Jun (cat)	160
Brown's Fishing (?) Excursion	Jun	320
The Curate's Dilemma	Jun	290
Home without Mother	Jun	220
Jam Making/from fruit picking to Chivers' factory	Jul 14	300
Jockeys vs Athletes/unusual cricket match	Jul 21	160
Bull Fight on Board Ship	Jul 28	220
*The Medium Exposed?Is Spiritualism a Fraud?	Jul	385
House to Let	Jul	
Just a Little Piece of Cloth	Jul	245

1906 *(continued)*

Title/description	Date	Length
Balloon Craze/balloon vs motor cars/Tedd to Cranleigh	Aug 4	165
The World's Wizard	Aug	350
Various Popular Liquors Illustrated	Aug	200
Spooning/facial of kissing couple	Aug	65
Barnet Horse Fair/incl Golding the human ostrich	Sep 8	140
Training a Horse for Steeple Chasing	Sep (cat)	370
Point to Point Racing	Sep (cat)	175
Oh That Doctor's Boy!	Sep	336
*Aberdeen University Quatercentenary Celebrations	Sep	2200
*The ? Motorist	Oct	190
Woman Supreme	Oct	212
Introductions Extraordinary	Nov	300
How to Make Time Fly/reversing clock	Nov	300
Jim the Signalman/son saves mother	Dec	354
The Lover's Predicament	Dec	245

1907

The Cook's Dream	Jan	320
A Tragedy of the Ice/"full of startling incidents"	Jan	95
The Chef's Revenge/trick	Feb	238
A Mother's Sin/maid abandons baby	Mar	356
A Knight Errant	Mar	480
His First Top Hat/Eton boy's new hat	Mar	260
*The Fatal Hand	Apr	415
The Bothered Bathers	May	275
Adventures of a Broker's Man	May	305
The Bookmaker/facial: watches race and welshes	May	140
The Burglar's Surprise	May	140
How a Burglar Feels/loots house then caught	May	198
An Inhuman Father/husband throws baby into lake	May	420
My Lady's Revenge/man fills rival, then killed	May	520
The Fidgety Fly/facial: man bothered by fly	May	248
The Tale of a Mouse	Jun	155
Sheep Shearing	Aug 29	230
The Cheaters Cheated	Aug	510
Pity the Poor Blind/facial: "blind" beggar button	Aug	107
The Amateur Paper Hanger	Sep	282
The Boy Bandits	Oct	480

1908

The Mauretania's First Voyage/launch etc	Jan 9	410
The Football Star/How Football Reports are Prepared	Nov 08/Jan 09	500
The Carter's Terrible Task/US: Great Trunk Mystery	Feb	419
The Magic Garden/trick	Apr	
The Prodigal Son/swineherd disillusioned	May	
The Tramp	Jul	
The Magic Box	Aug	

(continued)

1908 (continued)

Title/description	Date	Length
Wreck of the *Amazon*/off South Wales	Sep 10	
*Whaling Afloat and Ashore/Norwegian boats off Ireland	Sep 10	750
The Falls of Doonas/Irish scenery	Sep 17	
The Barry-Towns Sculling Race/river race	Oct 15	
Salmon Fishing/with Road and Fly	Oct 15	310
Village Life/scenes in Ireland	Dec 10	345
Lobster Fishing/coracle fishermen in NW Ireland	Dec 10	212
A Cattle Drive in County Galway/rustling reconstruction	Dec 24	274
The Robber and the Jew/Amateur thieves	Dec	320
The Lady Luna(tic)'s Hat/Lady blown to moon	Dec	1185

1909

The Newspaper World from Within/*Morning Leader*	Jan 21	1200
Domesticated Elephants/trained elephants perform	Jan 28	320
Love versus Science/Prof drugs wife's lover	Jan	500
My Dolly/Man disowns daughter	Jan	400
The Duped Othello/Theatrical Chimney Sweep	Jan	320
King and Queen's Visit to Berlin	Feb 11	
The Polite Parson	Mar	350
The Escapades of Teddy Bear/Taxidermist . . .	May	
*The Burning Home/Man sees child saved	May	485
Suspected; or, The Mysterious Lodger/Lodger's dummy	June	300
What Happened to Brown?	Sep	

1910

The Butterfly

A film attributed to Paul by several early writers, and now listed on IMDb, although not otherwise documented or known to exist. Possibly confused with James Williamson's *History of a Butterfly: A Romance of Insect Life* (1910).

Notes

Prelude

1. The start of Paul's demotion was Rachael Low's dismissive verdict in her and Roger Manvell's *History of the British Film*, vol. 1, *1896–1906* (London: Allen & Unwin, 1948), chap. 12. More recently, Richard Brown and Barry Anthony go to considerable lengths to impugn Paul's actions and accounts in *The Kinetoscope: A British History* (East Barnet: John Libbey, 2017).

2. Terry Ramsaye, *A Million and One Nights: A History of the Motion Picture through 1926* (New York: Simon & Schuster, 1926).

3. H. G. Wells, *When the Sleeper Wakes* (New York and London, 1899).

4. Low and Manvell, *History of the British Film*, 1:24.

5. John Barnes, *The Beginnings of the Cinema in England* (1976; Exeter: University of Exeter Press, 1998).

6. The entry on William Friese-Greene by Peter Domankiewicz and Stephen Herbert in *Who's Who of Victorian Cinema* succinctly explains the contradictions of his reputation. See http://www.victorian-cinema.net/friesegreene.

7. See my article "'Has the Cinema a Career?' Pictures and Prejudice: The Origins of British Resistance to Film," *Times Literary Supplement*, 17 November 1995, 22–23. This followed an earlier conversazione with John Carey at the British Academy, which considered the seemingly unique "English suspicion" of cinema.

8. Among the key books that signaled this new direction: Charles Musser, *Before the Nickelodeon: Edwin Porter and the Edison Manufacturing Company* (Berkeley: University of California Press, 1991) and *The Emergence of Cinema: The American Screen to 1907* (Berkeley: University of California Press, 1994); Richard Brown and Barry Anthony, *A Victorian Film Enterprise: The History of the British Mutoscope and Biograph Company, 1897–1915* (Trowbridge: Flicks Books, 1999).

9. As attempted in *The Last Machine*, the 1994 BBC television series and accompanying book, *The Last Machine: Early Cinema and the Birth of the Modern World* (London: BBC Educational Projects and BFI Publishing, 1994). Writing these first alerted me to the importance of Paul.

10. Despite the singular noun, audiences should always be considered plural and varied, calling for a variety of methodologies to probe their motives and composition. See my edited collection *Audiences: Defining and Researching Screen Entertainment Reception* (Amsterdam: Amsterdam University Press, 2012).

11. Luke McKernan's reflections on the growth of cinema and its audiences in London, resulting from his research for the AHRB London Project at Birkbeck College,

provide an invaluable proof of this proposition. See especially "Diverting Time: London's Cinemas and Their Audiences, 1906–1914," *London Journal* 32, no. 2 (July 2007): 125–44, and "A Fury for Seeing: Cinema, Audience and Leisure in London in 1913," *Early Popular Visual Culture* 6, no. 3 (2008): 271–80.

Chapter 1

1. Edison had not extended his patent beyond the continental USA, meaning that others were free to imitate or develop his device abroad. Paul was not the only pioneer to take inspiration from the Kinetoscope: Antoine Lumière saw it in Paris and urged his sons to improve on it, thus leading to the Cinématographe.

2. The key scientific papers—P. M. Roget, "Explanation of an Optical Deception in the Appearance of the Spokes of a Wheel Seen through Vertical Apertures" (Royal Society, 1824), and Michael Faraday, "On a Peculiar Class of Optical Deceptions" (Royal Institution of Great Britain, 1830)—influenced the development of the disk-based phenakistiscope, in 1832, by Joseph Plateau in Belgium and Simon Stampfer in Austria. A great variety of related disk and wheel devices followed throughout the nineteenth century, as authoritatively summarized in Wikipedia, https://en.wikipe dia.org/wiki/Phenakistiscope.

3. On the wide cultural influence of the Phonograph, see Ian Christie, "Early Phonograph Culture and Moving Pictures," in *The Sounds of Early Cinema*, ed. Richard Abel and Rick Altman (Bloomington: Indiana University Press, 2001), 3–12.

4. The phenakistiscope disk was developed by several scientists in the 1830s to exploit the illusion of sequential figures appearing in continuous motion, when viewed through slits in another disk—an early form of the intermittent motion and shutter that would be central to filmmaking and viewing. See Tom Gunning, "Doing for the Eye What the Phonograph Does for the Ear," in Abel and Altman, *Sounds of Early Cinema*, 13–31.

5. For the fullest account of Dickson's career, see Paul Spehr, *The Man Who Made Movies: W. K. L. Dickson* (New Barnet: John Libbey, 2008).

6. The London Stereoscopic Company, founded in 1854, offered a hundred thousand different stereographs, and it has been estimated that approximately five million stereographs were produced in the United States in 1864.

7. Oliver Wendell Holmes, "The Stereoscope and The Stereograph," *Atlantic Monthly*, no. 3 (June 1859), 738–48.

8. Charles Baudelaire, "The Modern Public and Photography," in "The Salon of 1859," *Révue Française* (Paris), 10 June–20 July 1859.

9. This hypothesis was proposed by Georges Potonniée, in *Les origines du cinématographe* (Paris: Publications Photographique Paul Montel, 1928). This was referenced by André Bazin in his well-known essay "The Myth of Total Cinema," unfortunately mistranslated in its most common English-language edition. See Christie, "Will the 3D Revolution Happen?" in *Techne*, ed. Annie van den Oever (Amsterdam: Amsterdam University Press, 2013), 124–26.

10. For a recent study of this neglected phase of coexistence, see Nicholas Hiley, "Lantern Showmen and Early Film," *Magic Lantern* (UK Magic Lantern Society), no. 15 (2018), 5–8.

11. Louis Lumière was strongly committed to achieving stereoscopic moving pictures and continued working on this until he succeeded in the mid-1930s.

12. See Michael Colmer, *The Great Vending Machine Book* (Chicago: Contemporary Books, 1977), 3. Also http://vendingmachine.umwblogs.org/antecedents/.

13. *Syracuse Daily Standard*, 25 December 1894, 3; *Middleton Daily Argus*, 20 March 1895, 7.

14. "I do not see there's anything to be made out of it. I have been largely influenced by sentiment in the prosecution of this design." "Thomas A. Edison: His Latest Invention," *Newark Daily Advocate*, 9 April 1894.

15. For instance, Reynaud's *théâtre optique* (patented 1888) and Marey's chronophotographic gun, described in *La Nature* in 1882. The intermittent illumination principle had already been used by Ottomar Anschütz in his Tachyscope of 1887. It has also been suggested that William Friese-Greene's 1889 patent for a "chronophotographic camera," a report of which from the *Photographic News* Friese-Greene sent to Edison, may have dissuaded him from seeking a British patent for the Kinetoscope.

16. A gross monthly revenue of $1,400 would result from 28,000 viewings at 5 cents each, or about 1,000 viewings per day.

17. On the detailed chronology of the Kinetoscope in Britain, see Richard Brown and Barry Anthony, *The Kinetoscope: A British History* (East Barnet: John Libbey, 2017)15–29.

18. His 1894 catalogue was named "Edison's Latest Wonders."

19. *Syracuse Daily Standard*, 25 December 1894, 3.

20. *Middletown Daily Argus*, 20 March 1895, 7; reprinted from the *Atlanta Constitution*.

21. *Coshocton* [Ohio] *Democratic Standard*, 21 December 1894; reprinted from the *New York Herald*.

22. "The Wonders of the Kinetoscope," *Bradford Daily Argus*, 20 December 1894, 3. Quoted by Richard Brown, "The Kinetoscope in Yorkshire," in *Visual Delights: Essays on the Popular and Projected Image in the 19th Century*, ed. Simon Popple and Vanessa Toulmin (Trowbridge: Flicks Books, 2000), 112–13.

23. Dickson, together with his sister Antonia, wrote and illustrated the promotional booklet *History of the Kinetograph, Kinetoscope and Kinetophonograph* (1895).

24. On Wells, see chap. 4, pp. 86–92. Among many references to optical devices in Proust, there is a direct reference to the Kinetoscope early in *Swanns's Way*: "We isolate, when we see a horse run, the successive positions shown to us by a kinetoscope" (Lydia Davis translation, 2002, p. 7). Other literary references include two by Jack London, in "Two Gold Bricks" (*The Owl*, 1897) and *Martin Eden* (1909).

25. Certainly in Australia, from the evidence of advertisements, and probably in many other countries remote from centers of equipment manufacture.

26. The Mutoscope was invented by Dickson, after he had left Edison in 1894, in partnership with Herman Casler, and led to the launch of the American Mutoscope Company the following year.

27. Handwritten notes by Acres in a copy of Frederick A. Talbot, *Moving Pictures: How They Are Made and Worked* (London: Heinemann, 1912), held by the British Film Institute.

28. Talbot, *Moving Pictures*, 33.

29. The cost of a Kinetoscope in America was $300, around £60 at the then exchange rate of 0.2; in Britain, the Continental Commerce Company charged £70, plus transport. Stating this price as upward of $8,000 in present-day terms is probably misleading, although buying Kinetoscopes was clearly a considerable investment. See

also Deac Rossell, *Living Pictures: The Origins of the Movies* (Albany: State University of New York Press, 1998); Ray Phillips, *Edison's Kinetoscope and Its Films: A History to 1896* (Westport, CT: Greenwood Press, 1997).

30. According to a favorable report in the electrical journal *Lightning*, noted by Brown and Anthony, *Kinetoscope*, 26.

31. Terry Ramsaye, *A Million and One Nights: A History of the Motion Picture through 1926* (New York: Simon & Schuster, 1926), 148.

32. Will Day, "R. W. Paul, MIEE 1894–1932," typescript, Day Papers 10, Bibliothèque de Film (BiFi), Paris.

33. *The Continental Commerce Company v. Georgiades and Others "Passing Off,"* heard in the High Court, January 1895. Discussed in Brown and Anthony, *Kinetoscope*, 28 ff.

34. Notice published in the *London Gazette*, 11 February 1895; reproduced in Brown and Anthony, *Kinetoscope*, 34. Although Brown and Anthony claim that Paul knew his actions were fraudulent and acted "defensively," having a contract with the American Kinetoscope Company would have been normal business practice in view of the resources committed.

35. According to John Barnes's research, Thomas Short was a partner in Short and Mason. Barnes, *The Beginnings of the Cinema in England* (1976; Exeter: University of Exeter Press, 1998), 1:126. Short and Mason, makers of barometers, anemometers, and compasses, were at 40 Hatton Garden, according to *Grace's Guide*.

36. Denis Gifford, *The British Film Catalogue*, vol. 1, *Fiction Film, 1895–1994*, 3rd ed. (London: Fitzroy Dearborn Publishers, 2000).

37. A number of Filoscopes survive and are found in various collections, including the Bill Douglas Collection at Exeter University and the National Media Museum in Bradford.

38. Irene Codd papers, British Film Institute, Special Collections, 48.

39. Information from Brown, based on the deposition for *Continental Commerce Company v. Georgiades.*

40. Kinetoscopes required substantial wooden cases, and it was customary for such woodwork to be contracted out to cabinetmakers. The father of Paul's future wife, Augustus Daws, was a master cabinetmaker, so might Robert and Ellen have first met as a result of this new direction in Paul's work?

41. "Before 1910," *Proceedings of the British Kinematograph Society*, no. 38 (1936), 2.

42. Acres had patented a device for rapidly changing slides in December 1893, as part of his interest in recording cloud formation and waves breaking on the seashore.

43. The case for Acres was made in a booklet by his grandson, Alan Birt Acres, *Frontiersman to Film-maker: The Biography of Film Pioneer Birt Acres, FRPS, FRMetS 1854–1918* (Projection Box, 2006). Research papers and recordings for this were deposited at the National Fairground and Circus Archive, University of Sheffield, see https://www.sheffield.ac.uk/nfca/collections/birtacres.

44. "Before 1910," 3.

45. Ramsaye, *Million and One Nights*, 149.

46. Birt Acres, "Amateur Photography," *Amateur Photographer*, 9 October 1896 (quoting from author's typescript).

47. This account, omitting all mention of Paul or their collaboration, appears in a website, *Birt Acres (1854–1918)—Film Pioneer*, https://wichm.home.xs4all.nl/acrestotal.htm.

48. Talbot, *Moving Pictures*, 37.

49. Day, "R. W. Paul" (BiFi), 2.

50. I am particularly grateful to Richard Brown for providing me with information about Acres's point of view in this dispute, which was long underrepresented; although I have not adopted it without putting Paul's side as well.

51. In March 1895, Paul was engaged in a battle of letters with his old employer Elliott Brothers over who was entitled to claim originality in the design of the Ayrton-Mather galvanometers in their respective catalogues. The early electrical business was as fraught with disputes over patents and priority as the film business.

52. Paul, quoted in Ramsaye, *Million and One Nights*, 151.

53. Letter from Paul to Messter, 5 August 1932; Bundesarchiv-Filmarchiv, Barnes Collection. Copies of this and subsequently cited letters kindly supplied by John Barnes.

54. Messter's reference was to an early, and much criticized, reference work, Henry V. Hopwood, *Living Pictures: Their History, Photo-Production and Working* (London: Hatton Press, 1915)

55. This is based on correspondence with Brown, and reflects the view expressed in his articles and books. It may be worth noting that the Lumières made use of the specialist engineer Jules Carpentier to manufacture the Cinématographe. See Laurent Mannoni on Carpentier in *Who's Who of Victorian Cinema*, http://www.victorian-cinema.net/carpentier.

56. Acres, "Amateur Photography" (quoting typescript; see note 46).

57. Rossell, *Living Pictures*, 92.

58. This is also known as a "Geneva mechanism," after its creation by Swiss watchmakers. See, for instance, John H. Bickford, "Geneva Mechanisms," *Mechanisms for Intermittent Motion* (New York: Industrial Press, 1972), 128.

59. Acres, *Frontiersman to Filmmaker*.

60. On Muybridge's biography and achievements, see Marta Braun, *Eadweard Muybridge* (Chicago: University of Chicago Press, 2010).

61. Paul's 1914 speech to his employees outlined his expectations in uncompromising terms (see chap. 10, pp. 208–10). A memoir by the future film entrepreneur Albany Ward, who worked for Acres, painted him as demanding.

62. In the case of Dickson's film, the fact that two men are shown dancing has attracted speculation, despite the fact that this was not uncommon in all-male work situations in Victorian and Edwardian times. See, for instance, Paul's 1908 film *Whaling Afloat and Ashore*, which includes men dancing together at a remote island whaling station.

63. Report of the Kinematograph Manufacturers Association annual dinner, which Paul had chaired. "Our Trade Dinner," *Kinematograph and Lantern Weekly*, 1908 (hereafter cited as *Kine Weekly*).

64. I am grateful to Richard Brown for confirming that this note appeared in Acres's hand on p. 38.

65. See chap. 11 for further discussion of the story's legacy.

66. Acres, "Amateur Photography" (typescript).

67. See, for instance, Jay David Bolter and Richard Grusin's "remediation" thesis on how new visual media achieve cultural significance and validation by refashioning such earlier media achievements. Bolter and Grusin, *Remediation: Understanding New Media* (Cambridge, MA: MIT Press, 1998). This explicitly invoked Marshall McLuhan's and Harold Innes's theses on new media encapsulating older ones.

68. Appearing, for instance, in the secondhand sales list of Harrington and Co., Sydney and Brisbane, *Australian Photographic Journal* (Sydney), April–September 1899, xix, xxi.

69. Composers of the nineteenth and early twentieth centuries were also fascinated by the challenge of evoking the sea's changing moods, from Mendelssohn and Wagner to Debussy, and in Britain, Bridge, Vaughan Williams, Britten, and many others.

70. This was "discovered" as recently as 2009, when the George Williams collection of fairground films from 1896 was donated to the National Fairground and Circus Archive, Sheffield University. See https://www.sheffield.ac.uk/nfca/collections/geor gewilliams. Another "crime drama," *Footpads*, has long been attributed to Paul-Acres, appearing both on the BFI website and on the DVD that I compiled. But since it appears in no contemporary listing, and makes use of a very different setting—which appears to be a theater backdrop—there seems no justifiable reason to attribute it to Paul or Acres in 1896.

71. See chap. 3, pp. 65–66.

72. Brown and Anthony scorn this venue as a notorious address with a show fronted by a bogus "Viscount," but admit that Paul's exhibition was probably a "trial run" for the ambitious Earls Court installation ten days later. Brown and Anthony, *Kinetoscope*, 121–27.

73. *English Mechanic*, 5 April 1895. In the same issue Cecil Wray of Bradford offered a "list of 50 subjects" for the Kinetoscope, which included both Acres-Paul and Edison titles—although it is not clear if his distribution of the latter was authorized by the Edison Manufacturing Company. See Brown, "Kinetoscope in Yorkshire," 108.

74. Brown and Anthony note that Paul was offering Kinetoscopes for sale as late as October 1897, in order to exhibit the Jubilee procession as "a complete show in itself," apparently still having surplus machines on his hands. Brown and Anthony, *Kinetoscope*, 89.

75. Paul was selling Kinetoscopes for £40. See Richard Brown, "A New Look at Old History—the Kinetoscope: Fraud and Market Development in Britain in 1895," *Early Popular and Visual Culture* 10, no. 4 (November 2012): 417.

76. For an 1882 Boston poster for *Around the World*, see https://www.loc.gov /item/2014636784/.

77. A number of Korda's productions of the 1930s were set in different periods and parts of the British Empire: *Sanders of the River* (1935, set in modern Nigeria), *Fire over England* (1937, stressing Elizabeth's territorial ambitions), *The Drum* (1938, set during the Indian Raj), *The Four Feathers* (1939, set in 1880s Sudan), and more remotely *That Hamilton Woman* (1941) and *The Jungle Book* (1942). As he said during the late 1930s, "an outsider often makes the best job of a national film." Quoted in Karol Kulik, *Alexander Korda: The Man Who Could Work Miracles* (London: W. H. Allen, 1975), 97.

78. *Times*, 18 April 1896.

79. "Before 1910," 2.

80. See chap. 4.

Chapter 2

1. Acres did not return to exhibition after his venue was closed by fire in 1896. And although the Lumières developed a worldwide network of franchise-holders and

roving operator-projectionists between 1896 and 1898, the brothers largely withdrew from the motion-picture business after 1900, preferring to work on such new projects as color photography and stereoscopy rather then become involved with "theatrical" production, as Paul did.

2. Irene Codd papers, British Film Institute, Special Collections. The papers, deposited at BFI in 2014, comprise a much-revised typescript of about eighty pages; a letter from "Ellen and Robert Paul" to Codd, dated 25 April 1938; and various other letters and newspaper cuttings. I am grateful to Bryony Dixon and Nathalie Morris for the opportunity to consult and quote from this.

3. Irene's father, Mortimer Arthur Codd (1880–1938), was an electrical engineer and scientist, director of research at the X-Ray Laboratories of Torrington Place in the 1920s, and, like Paul, a motoring enthusiast who published various books on electrical systems for motor cars.

4. Paul's birthplace was wrongly given by John Barnes as Albion Place, when in fact it was Albion Road (now Furlong Road), off Holloway Road. The artist and photographer Sam Nightingale has investigated changing street names, trying, like myself, to establish exactly where Paul was born, as part of his "lost cinemas" project. https://www.samnightingale.com/launched-islingtons-lost-cinemas-online-archive/.

5. In the 1899 Lloyd's Register, the following ships are listed as owned by Paul and Shellshear of 21 Billiter Street, London EC1: *Bolderaa, Neva, Octa, Onega, Varna, Viatka*. By 1906, only the *Bolderaa* was listed. I am grateful for help in tracing these ships from marine historian Gilbert Provost.

6. "The Incorporation of West Ham," *Times*, 1 November 1886, 12.

7. Robert's "home address" was given in the Finsbury Technical College Record as "Frith Knowl, Elstree, Herts," after 1887.

8. G. Watchorn, *Never a Dull Moment: A History of the Cambridge Instrument Company 1918–1968* (Cambridge, 1990), 3.

9. A. E. Douglas-Smith, *The City of London School* (Oxford: Basil Blackwell, 1937).

10. Finsbury Technical College and Old Students Association, *Journal of the City and Guilds of London Institute*, n.s., 1, no. 2 (March 1888): 66. I am grateful to John Barnes for this reference.

11. The first cable broke after two months and 732 messages successfully sent.

12. See chap. 11, p. 219.

13. Matthew Josephson, *Edison: A Biography* (New York: McGraw-Hill, 1959), 252.

14. From the account by Robert Harding in the William J. Hammer Collection, Smithsonian Institution, Washington, DC, http://americanhistory.si.edu/archives /d8069.htm (accessed 12 December 2006). This collection, from one of Edison's employees, includes photographs of the Edison dynamo and the 1882 electric lighting plant erected by Hammer. The official catalogue of the International Electric and Gas Exhibition and articles from the *Daily Telegraph*, *Daily Chronicle*, and *Daily News* are also included (series 4, box 99; series 3, box 42, folders 1–2).

15. Hugo Hirst, "The History of the General Electric Company up to 1900, Part 1," *GEC Review* 14, no. 1 (1999): 48.

16. Hirst, "History of the General Electric Company," 50.

17. Hannah Gay, "East End, West End: Science, Education and Culture in Victorian London," *Canadian Journal of History* 32 (August 1997): 159.

18. John Scott Russell (1808–1882), the engineer and naval architect, is quoted in D. S. L. Cardwell, *The Organisation of Science in England* (London: Heinemann, 1972),

111–12. Christopher Keep, "Technology, Industrial Conflict and the Development of Technical Education in 19th-Century England," *Victorian Studies* 47, no. 2 (Winter 2005): 299.

19. James Foreman-Peck, "Spontaneous Disorder? A Very Short History of British Vocational Education and Training," *Policy Futures in Education* 2, no. 1 (2004): 81.

20. The speech, at a prizegiving in Greenwich for the Government Science and Arts examinations in 1875, was made while Gladstone was in opposition after his first ministry. Quoted in Hannah Gay, *The History of Imperial College London, 1907–2007: Higher Education and Research in Science, Technology and Medicine* (London: Imperial College Press, 2007), 27–28.

21. Gay, *History of Imperial College London*.

22. Information on William Taylor from *Obituary Notices of Fellows of the Royal Society* 2 (1937): 363.

23. Hirst, "History of the General Electric Company," 52.

24. See H. Bowden, seminar on the history of electrical engineering, "The Test Room," Institute of Electrical Engineering 2003/10330, *IEE Digest*, no. 4 (2003).

25. The company had been set up by Francis Welle, with the objectives of "the production, sale, purchase, and rental of telephone and telegraph equipment and everything that has to do directly or indirectly with the use of electricity." By the turn of the century it had grown enormously, and as the European base of ITT, would employ eleven thousand by 1927. See "Inside Alcatel-Lucent: The Company That Makes the Web," https://www.ft.com/content/4e7d2682-33bd-11e4-ba62-00144feabdc0 (accessed 12 April 2019).

26. The execution of Leon Czolgosz, the anarchist responsible for assassinating President William McKinley in 1901, was re-created for Edison's series of films on McKinley's death and its aftermath. Edison was personally involved in the preparation for *Electrocuting an Elephant* (1903), which showed Topsy, an elephant condemned to death after injuring various spectators at Luna Park, succumbing to a lethal dose of alternating current. This spectacular demonstration formed part of Edison's campaign to show that AC was more dangerous than his DC system.

27. That Tesla (1846–1943), born in what is now Croatia and naturalized as American, died in poverty has helped to cast him as a victim. So too has the long debate over priority in wireless invention. Tesla's original patents of 1900 were challenged by Marconi, but the initial judgment in Marconi's favor was reversed by the US Supreme Court in 1943, shortly after Tesla's death.

28. "Before 1910," *Proceedings of the British Kinematograph Society*, no. 38 (1936), 2.

29. On polemics during this period and new requirements in current measurement, see Graeme J. N. Gooday, *The Morals of Measurement: Accuracy, Irony, and Trust in Late Victorian Electrical Practice* (Cambridge: Cambridge University Press, 2004), 155 ff.

30. C. N. Brown, "R. W. Paul and the Unipivot Galvanometer," IEE Seminars on the History of Electrical Engineering, *IEE Digest*, no. 3 (2003), p. 3.

31. Crookes's talk, "The Mechanical Action of Light," was accompanied by a demonstration of his radiometer and attracted Thompson's interest to the study of light.

32. Quoted in A. C. Lynch, "Sylvanus Thompson: Teacher, Researcher, Historian," *Institute of Electrical Engineering Proceedings* 136, pt. A, no. 6 (November 1989): 307.

33. Hirst, "History of the General Electric Company," 50.

34. The Society of Telegraph Engineers, founded in 1874, changed its name to

reflect a widening electrical profession on 1 January 1889. Associates had to be twenty-one, which would just have made Paul eligible. See "Research Guide: History of the Institution," Institute of Engineering and Technology (IET) website, http://www.iee .org/TheIEE/Research/Archives/ResearchGuides/IEE&SavoyPlace.cfm (accessed 14 December 2006). Richard Brown notes that another future Kinetoscope dealer, Cecil Wray of Bradford, also became an associate of the IEE in 1889. See Brown, "The Kinetoscope in Yorkshire," in *Visual Delights: Essays on the Popular and Projected Image in the 19th Century*, ed. Simon Popple and Vanessa Toulmin (Trowbridge: Flicks Books, 2000), 108. Brown later assumes that Paul was "upper middle-class" (113), but this seems unwarranted on the evidence currently available—or requires a more precise historical definition of "middle class."

35. Lynch, "Sylvanus Thompson," 309.

36. Richard Brown has identified seven patent applications made by Paul between 1887 (when he was only eighteen), and his Kinetoscope Apparatus application of September 1895.

37. Thompson gave the Christmas Lectures in 1896 on "Light: Visible and Invisible" and in 1910 on "Sound: Musical and Non-Musical."

38. Paul, "Before 1910," 6. This would probably have been in connection with Thompson's successful book *Elementary Lessons on Electricity and Magnetism*, first published 1881, updated and in print until 1920.

39. C. V. Boys, *Soap Bubbles: Their Colours and the Forces Which Mould Them* (New York: SPCK, 1912); A. M. Worthington, *A Study of Splashes* (London: Longman, Green & Co, 1908). Both have been reprinted by Dover.

40. Such documentation began early in Boys's case, with photographs published in William G. FitzGerald, "Some Curiosities of Modern Photography II," *Strand*, no. 191 (February 1895).

41. IEE register 1896.

42. Acres, 1912 annotations to Frederick A. Talbot, *Moving Pictures: How They Are Made and Worked* (London: Heinemann, 1912), 43. The Cinématographe was first demonstrated publicly in Paris at a photographic trade event on 22 March 1895. See Deac Rossell, *Living Pictures: The Origins of the Movies* (Albany: State University of New York Press, 1998), 130–31.

43. The origin of the projecting lantern has traditionally been attributed to Kircher, on the basis of his book *Ars Magna lucis et umbrae* (The great science of light and dark), first published in Rome in 1646. However, inconsistencies in Kircher's account have suggested he did not in fact know how to make a lantern, and that the credit for this should go to Huygens, who mentioned the "lanterne magique" in a manuscript of 1659 (*Oeuvres completes*, vol. 22), which included nine sketches of its arrangement, together with typical skeleton slide images. *Encyclopedia of the Magic Lantern* (Magic Lantern Society, 2001), 142. See also http://www.luikerwaal.com /newframe_uk.htm?/kircher_uk.htm .

44. Limelight was produced by an oxy-hydrogen-powered flame playing on a block of lime. Double and triple lanterns allowed time-lapse "dissolves" and motion effects to be displayed.

45. Report of an ordinary meeting of the Royal Photographic Society, *Photographic Journal*, 31 January 1896, 123–24.

46. Paul, letter to Messter, 5 August 1932; Bundesarchiv-Filmarchiv, Barnes Collection. See chap. 11 for details of the Paul-Messter correspondence.

47. The main "discourse" on 28 February was by a Dr. John Murray, with "marine organisms shown under microscopes." Friese-Greene exhibited a "cylindrical photo printing machine," while Paul showed the Theatrograph and "electrical instruments." Richard Brown, who kindly provided this information, was informed by the Institution's archivist that the attendance of 320 for Murray's presentation was "relatively low."

48. McLeod's diary notes paying 5 pounds for an Ayrton-Mather model galvanometer on 26 June 1894, purchased on behalf of the British Association Solar Spectroscopy Committee. In am grateful to Hannah Gay for this information.

49. *English Mechanic*, 21 February 1896, 11.

50. S. P. Thompson, "The Organisation of Secondary and Technical Education in London," *Journal of the Society of Arts*, no. 38 (1890), 324. Quoted in A. C. Lynch, "Sylvanus Thompson," 308.

51. The shows staged by Sir Augustus Henry Glossop Harris (1852–1896), nicknamed the "Father of Modern Pantomime" or "Augustus Druriolanus," were hugely successful, although criticized by some for vulgarity. While he did not invent the conventions of Principal Boys being played by women and Dames played by men, Harris developed the tradition, with popular music-hall stars like Dan Leno playing the Dame every year. See Victoria and Albert Museum, Theatre collections, http://www.vam.ac.uk/content/articles/d/dan-leno/ (accessed 10 April 2019).

52. Terry Ramsaye, *A Million and One Nights: A History of the Motion Picture through 1926* (New York: Simon & Schuster, 1926), 238–89; Leslie Wood, *The Miracle of the Movies* (London: Burke Publishing, 1947), 100–101.

53. According to Ramsaye (*Million and One Nights*, 239), who would presumably have been informed of this by Paul.

54. Advertised in *The Era*, 21 March 1896, 16.

55. Maskelyne was listed as a sponsor of Paul's proposed public flotation in 1897, and one of Paul's products was named after him.

56. Philippe De Loutherbourg was the first to advertise his show as "moving pictures," in 1781; see Ann Bermingham, "Technologies of Illusion: De Loutherbourg's Eidophusikon in Eighteenth-Century London," *Art History* 2 (April 2016): 376–99. On the 1930s "news theater" trend in cinema exhibition, see http://bufvc.ac.uk/newsonscreen/davidlean/news-theatres.

57. Richard Brown has discovered that Paul had approached other theaters before joining the Alhambra program. Although he remembered the engagement running for over two years, he actually left the Alhambra in autumn 1897, after a dispute about the terms for showing his films of Queen Victoria's Diamond Jubilee.

58. Wood, *Miracle of the Movies*, 101, purporting to quote from interview with Paul.

59. Wood, *Miracle of the Movies*, 102.

60. Ramsaye, *Million and One Nights*, 240–41.

61. Ramsaye, *Million and One Nights*, 240–41.

62. Wood, *Miracle of the Movies*, 101.

63. A review of Paul's Theatrograph show at the Royal Institution remarked that "in its present condition the registration of the pictures on the screen is not so perfect as in Lumière's arrangement." *Amateur Photographer* 23, no. 597 (13 March 1896): 226.

64. Paul, letter to Messter, 5 August 1932, 3.

65. John Barnes, "Robert William Paul, Father of the British Film Industry," *1895* (journal of the Association française de recherche sur l'histoire du cinéma [AFRHC]), no. 24 (June 1998), 6.

66. Paul, letter to Messter, 30 January 1935, 2.

67. Paul, letter to Messter, 30 January 1935, 2.

68. Carl Hertz, *A Modern Mystery Merchant* (London: Hutchinson,1924), 140.

69. Will Day, "R. W. Paul MIEE," typescript, Day Papers 10, Bibliothèque du Film (BiFi), Paris, 6.

70. Paul, letter to Messter, 5 August 1932.

71. This first improvised camera of Méliès is preserved in the Cinémathèque Française collection.

72. Charles Pathé, founder along with his brothers of the company bearing his name, was a Phonograph exhibitor before he moved into Kinetoscopes and then film production and distribution. In December 1897, the company was recapitalized as Société Pathé Frères and began operating on a similar level to Paul.

73. An exchange in the *British Journal of Photography* was eventually terminated by the editor.

74. Acres, letter in *British Journal of Photography*, 7 March 1896, quoted by Barnes, *Cinema in England*, 1:35. William Friese-Greene had long claimed priority in moving pictures on the basis of his 1889 patent for a "chronophotographic" camera, reported in *Photographic News*, 20 February 1890, and in *Scientific American*, 19 April 1890.

75. *Optical and Magic Lantern Journal* 7, no. 86 (July 1896): 119.

76. See the section "Paul's Afterlife" in chap. 11.

Chapter 3

1. The receipts for one day at Olympia in April 1896 were given as £63. 16s in the audited flotation prospectus of March 1897, published in a number of papers, including *The Era* and *Bristol Times and Mirror*, 24 April 1897.

2. "Animatograph" was spelled both without and with a final *e*, and although initially coined to differentiate Paul's Alhambra show, gradually eclipsed his original projector name of Theatrograph.

3. Edison would concentrate on selling equipment and building up a large catalogue of films. The Lumière catalogue eventually numbered over two thousand titles, and their original show at the Salon Indien in Paris would run for an unparalleled six years, until 1901. See Jean-Jacques Meusy, *Paris-Palaces ou le temps des cinémas (1894–1918)* (Paris, CNRS Editions, 1995; 2nd ed., 2002), appendix "Les cinémas et les lieux de projection parisiens jusqu'en 1918," 531.

4. "Before 1910," *Proceedings of the British Kinematograph Society*, no. 38 (1936), 4. The original film, thought to have survived only as a Filoscope published by Henry Short, has been restored by the Cineteca Nazionale in Rome from additional material found in the Portuguese archive and in the Bradford National Science and Media Museum.

5. See a playbill advertising *A Soldier's Courtship* between *The Wedding Gown* and *My Neighbour's Wife Tit for Tat*, as presented at the New Theatre, Bridgnorth, Shropshire, in 1834, in the playbill collection of the Templeman Library, University of Kent, UKC/POS/BDG N: 0593986. See also "Nineteenth-Century Drama," in the *Cambridge History of English and American Literature*, vol. 13.

6. Leslie Wood, *The Miracle of the Movies* (London: Burke Publishing, 1947), 102.

7. Wood's account refers to extra "players [who] appeared as atmosphere—people strolling in the park," but the film as now restored has no such extras, or indeed the lamppost mentioned by Wood. Had he embellished Paul's account, or had Paul himself misremembered, perhaps confusing it with his 1898 remake? The original film would have been unavailable for viewing by 1937.

8. Wood, *Miracle of the Movies*.

9. On rival US "passion play" films in 1897–1898, see Charles Musser, *The Emergence of Cinema: The American Screen to 1907* (Berkeley: University of California Press, 1994), 208–19.

10. It was the availability of celluloid strips coated with photographic emulsion that had made possible the burst of moving-picture activity in 1895–1896. These were initially supplied by the American Blair Camera company, which had a "European" branch in England from 1893. However, pioneer filmmakers generally preferred Eastman stock, imported from Rochester, New York. See John Barnes, *The Beginnings of the Cinema in England* (1976; Exeter: University of Exeter Press, 1998), 1:218–22.

11. A restoration of *The Soldier's Courtship*, based on scattered fragments in Rome, Lisbon, and Bradford, was commissioned by the Cineteca Nazionale in Rome from the German company Omnimago in 2011.

12. This film, currently lost, has been confused with both Acres's *Arrest of a Pickpocket* (rediscovered, and shown in Pordenone in 2006) and a film known as *Footpads*, which was long attributed to Paul, although it does not appear in any of his lists. The IMDb listing for *Arrest of a Bookmaker* shows a still from *Footpads*.

13. An Edison film of 1897, *Arrest in Chinatown San Francisco*, showed two Chinese criminals being forcibly led toward the camera, although this appears an exception. Charles Musser, *Edison Motion Pictures* (Pordenone: Le Giornate del Cinema Muto, 1997): 335. Otherwise, it is not until Wallace McCutcheon's *How They Rob Men in Chicago* (1900) and *The Dude and the Burglars* (1903), both for American Mutoscope & Biograph, that violence appears in American films. In France, there is a rare example of early violence in Méliès's *Pickpocket et policeman* (1899).

14. Gifford, *British Film Catalogue*, 1:4.

15. Rachael Low and Roger Manvell, *The History of the British Film*, vol. 1, *1896–1906* (London: Allen & Unwin, 1948), 86. See chap. 12, "Paul and History."

16. Paul Langford, *Englishness Identified: Manners and Characters, 1650–1850* (Oxford: Oxford University Press, 2000). See chapter 3, "Decency," with its first section titled "Barbarity."

17. Paul Langford, *Englishness Identified: Manners and Character, 1650–1850* (Oxford: Oxford University Press, 2001), 145. The Defoe quotation is from his picaresque crime novel *The History and Remarkable Life of the Truly Honourable Col. Jaque* (1772).

18. Generally believed by film historians to have provided the template for Porter's celebrated *Great Train Robbery* (1903), made later in the same year after Mottershaw's film was bought for America.

19. *Prince and Princes of Wales Arriving in State at the Cardiff Exhibition* (Northern Photographic Works [Acres], 40 ft). Gifford, *British Film Catalogue*, 2:6.

20. The official website of the British monarchy, http://www.royal.gov.uk/LatestNewsandDiary/Pressreleases/2002/TheRoyalFilmPerformance2002.aspx.

21. Leslie Wood, *Romance of the Movies* (London: Heinemann, 1937), 39–40.

22. *Sportsman*, 5 June 1896, quoted in Will Day, "R. W. Paul, MIEE 1894–1932," typescript, Day Papers 10, Bibliothèque du Film (BiFi), Paris, 7. The writer alludes to the anecdote about George Washington, having attacking his father's tree with a hatchet, declaring when questioned that he could not tell a lie.

23. *Strand*, August 1896, 140; report also in *The Era*, 6 June 1896, 16.

24. Despite film scholars' refutations, the popular belief that audiences were scared by the Lumière film remains entrenched. See, for instance, Martin Loiperdinger and Bernd Elzer, "Lumière's Arrival of the Train: Cinema's Founding Myth," *Moving Image* 4, no. 1 (Spring 2004): 89–118.

25. *Ben-Hur* opened on Broadway in 1899 and appeared at the Theatre Royal, Drury Lane, in London in 1902, with a spectacular chariot race as its climax, using live horses on a treadmill, in front of a massive cyclorama. *The Whip* followed in 1909, with even more horses, simulating the Newmarket Gold Cup.

26. Irene Codd's memoir claims that the Prince of Wales had already seen Ellen Daws dance at the Alhambra, and that she was asked by the manager to be on hand when the prince came to see Paul's film.

27. The coronation of Victoria's nephew, Tsar Nicholas II, in May 1896 had been filmed by Lumière operators, with several short scenes of the procession entering and leaving the Cathedral of the Assumption in the Kremlin being widely shown. However, the presentation of the tsar to his subjects at Khodynka Field four days later turned into a disaster when the crowds stampeded, and Lumière film of the event was impounded. Jay Leyda, *Kino: A History of the Russian and Soviet Film* (London: George Allen & Unwin, 1960), 19–20.

28. David Brewster, *Letters on Natural Magic* (London: John Murray, 1834). On the history of "philosophical toys," automata, see my chapter "Toys, Instruments, Machines: Why the Hardware Matters," in *Multimedia Histories: From the Magic Lantern to the Internet*, ed. James Lyons and John Plunkett (Exeter: University of Exeter Press, 2007), 3–17.

29. David Devant, "Illusion and Disillusion," *Windsor Magazine*, December 1935, 117–26.

30. John Barnes reproduced an advertisement for the Devant interval presentation on 16 September 1896 as a supplement to volume 1 of the 1998 edition of *Cinema in England*, adding that this "continued for several years" in the upstairs Small Hall, seating five hundred, with an admission price of 6d. In January 1902, Paul's *Scrooge* "filled the whole of the interval without the need for any supporting films." The Queen's Hall, where the Promenade Concerts originated in 1895, was destroyed by bombing during World War II, leading to the concerts being moved to the Royal Albert Hall.

31. "Unsolicited Testimonial from David Devant, Esq.," in Robert W. Paul, *Illustrated Catalogue of Animated Photograph Films* (August 1898), list no. 15.

32. At this period, Paul's film did not have formal titles, but were identified for ordering purposes by single code-words. *Maskelyne* is not known to have survived, although there is a still.

33. The Davenports were "exposed" by Robert-Houdin and many others, who demonstrated how they achieved their supernatural effects. Even the showman P. T. Barnum denounced them as frauds in his book *The Humbugs of the World* (New York: Carleton, 1865). Maskelyne became a professional magician largely due to his cam-

paign against the Davenports and other professed mediums, publishing two books as part of his continuing campaign: *Modern Spiritualism* and *The Fraud of Modern Theosophy* (both London: Frederick Warne, 1875).

34. The recent discovery of a photograph of Booth has enabled him to be identified in a number of Paul's films, although this does not constitute evidence of him being the "director." See chap. 12, pp. 245–46.

35. Nevil Maskelyne and David Devant, *Our Magic: The Art and Practice of Magic* (London: G. Routledge, 1911), 166.

36. Maskelyne and Devant, *Our Magic*, 174–76.

37. Maskelyne and Devant, *Our Magic*, viii.

38. Day, "R. W. Paul" (BiFi), 6.

39. Allowing time between films, which would often be filled by a lantern slide and, no doubt, suitable patter by the presenter.

40. The late David Williams kindly supplied this information in "Leicester's First Film Show," an addendum to his book *Cinema in Leicester 1896–1931* (Leicester: Heart of England Press, 1993).

41. Williams makes this point in "Leicester's First Film Show." *The Boxing Kangaroo* was a Paul-Acres subject from 1895, probably already seen in Leicester on the Kinetoscope, while Paul appears to have duplicated various Edison subjects, which may have been the source of the "styles of dance."

42. The subtitle of Felix Mesguich's memoir *Tours de manivelle* was *Souvenirs d'un chasseur des images* (*Turning the Handle Memoirs of an Image Hunter*) (Paris: Grasset, 1933). The Lumière catalogue was extensively subdivided into *vues* of different types, including *vues exotiques* and various country and city groups from around the world.

43. *The Clock Tower, London Road, Pantomime Rehearsal*, in Paul, *Illustrated Catalogue*, list no. 15, p. 30.

44. *Leicester Journal*, 28 December 1896; *Leicester Daily Post*, 22 December 1896. I am grateful to David Williams for these and other references, from Leicester and Durham.

45. *Leicester Daily Post*, 28 December 1897.

46. Frank Gray, "The Sensation of the Century: Robert Paul and Film Exhibition in Brighton in 1896/97," in *Visual Delights—Two: Exhibition and Reception*, ed. Vanessa Toulmin and Simon Popple (Eastleigh: John Libbey, 2005), 219–35.

47. The film is not known to exist. What was previously believed to be *Scene on Brighton Beach*, held by the BFI National Archive, is more likely a film by the Brighton filmmaker Esme Collings.

48. The Douglas group totaled six films, none of which are known to survive, and included a *Scene on Douglas Beach*, which suggests the use of the same formula. Paul, *Illustrated Catalogue*, 30.

49. Paul's memoir appeared in 1936 in the *Journal of the Society of Motion Picture Engineers* under the title "Kinematographic Experiences," with a number of illustrations, including one of his shows at the Cheltenham Cricket Club on 4 and 5 December 1896. Two illustrations in volume 1 of Low and Manvell, *History* (figs. 9, 11). showed Cheltenham programs from 1896 and 1898.

50. *The Looker-On* and *Cheltenham Examiner*, December 1896. I owe these references to Patricia Cook, from her research on the Cheltenham context, as part of her PhD on the itinerant exhibitor Walter Slade (Birkbeck College, 2016). The sea cave was here misdescribed as being in Galicia, Spain.

51. The cities reached included Edinburgh, Glasgow, Sheffield, Burnley, Hull, Liverpool, Cardiff, and others not yet identified.

52. *Times*, 4 June 1895, 8.

53. Paul advertised a group of four films, including *The Soldier's Courtship*, in *The Era*, 16 May 1896, 16. References and advertisements appeared in the *Showman* from early 1901, including an important notice of "Films Let Out for Hire" in January 1902. On the *Showman* and film advertising, see Vanessa Toulmin, "'Local Films for Local People': Travelling Showmen and the Commissioning of Local Films in Great Britain, 1900–1902," *Film History* 13 (2001): 121–33.

54. Vanessa Toulmin, *Electric Edwardians: The Story of the Mitchell and Kenyon Collection* (London: British Film Institute, 2006), 16.

55. Quoted by Will Day, "An Early Chapter of Kinematograph History," *Kinematograph Year Book*, 1915, p. 47.

56. Letter from Jack Smith, Manager, Animatographe Sales Department, to S. H. Carter, Bradford, dated 16 September 1902; kindly made available by the National Fairground and Circus Archive, University of Sheffield. Paul's letterhead at this time advertises him as "Patentee and Manufacturor of the 'New Century' Animatograph," which might have been the source of Carter's company's name.

57. Reproduced in Toulmin, *Electric Edwardians*, 16.

58. See chap. 2, pp. 38–39.

59. Review in *The Era*. Frank Matcham (1854–1920) would later design the Victoria Palace, the London Coliseum, the London Palladium, the Hackney Empire and many theaters outside London.

60. See Begona Soto, "Rethinking Boundaries: The First Moving Images between Spain and Portugal," in *Networks of Early Film Distribution 1895–1915*, ed. Frank Kessler and Nanna Verhoeff (Eastleigh: John Libbey, 2007): 7–15.

61. One of the most famous of international lantern lecturers, Burton Holmes, coined the term "travelogue" for his presentations and was one of the first to add films to his repertoire. See Theodore X. Barber, "The Roots of Travel Cinema: John L. Stoddard, E. Burton Holmes and the Nineteenth-Century Travel Lecture," *Film History* 5, no. 1 (March 1993): 68–84. Picture postcards appeared in the 1880s and soon began to define the international iconography of "sight-seeing." See Elizabeth Edwards, *The Tourist Image: Myths and Myth Making in Tourism* (Chichester; John Wiley and Sons, 1996).

62. At least four other films were shot by Short; see listing in Barnes, *Cinema in England*, 1:258.

63. Advertisement in the *Magic Lantern Journal Annual, 1897–1898* (October 1897), xcv.

64. Advertisement for Animated Photograph films, *Photography Annual for 1897*, 1.

65. Paul, "Before 1910." See chap. 7 for further discussion of the Spanish and Portuguese and Swedish series.

66. Advertisements in *Aftonbladet*, June–August 1897. I am grateful to Christopher Nitzen in Stockholm, and to Karen Brookfield for translations of these.

Chapter 4

1. See text in appendix A.

2. See chap. 11, p. 222, and epilogue, pp. 256–57.

3. "The Prince's Derby: Shown by Lightning Photography," *Strand* 12 (August 1896): 134.

4. *The Era*, 25 April 1896, 17.

5. *Amateur Photographer*, 29 October 1895.

6. Terry Ramsaye, *A Million and One Nights: A History of the Motion Picture through 1926* (New York: Simon & Schuster, 1926), 160.

7. See, for instance, Olive Cook's pioneering *Movement in Two Dimensions* (London: Hutchinson, 1963); C. W. Ceram, *The Archaeology of Cinema* (London: Thames and Hudson, 1965); and the chronologies of most museums and displays dealing with the history of cinema.

8. J. W. Dunne, previously a pioneer aircraft designer, believed that "pre-cognitive" dreams could foretell future events, with implications of our understanding of time. His *An Experiment with Time* (1927) enjoyed wide currency during the 1930s, especially influencing J. B. Priestley's "time plays." See also my discussion of "time culture" in Britain between the wars, in *A Matter of Life and Death* (London: British Film Institute, 2000), 25-28.

9. See Norman MacKenzie and Jeanne MacKenzie, *The Time Traveller: The Life of H. G. Wells* (London: Weidenfield and Nicolson, 1963), 107. The "last man" is probably a reference to Mary Shelley's postapocalyptic novel of that title. Wells later cut the passages included at Henley's urging.

10. Day, "R. W. Paul" (BiFi), 10.

11. The Normal School became the Royal College of Science in 1890, and in 1907 formed part of the Imperial College of Science and Technology.

12. Ramsaye, *Million and One Nights*, 153.

13. Ramsaye, *Million and One Nights*, 159-60. Flammarion's *Lumen* was published in 1872, in a series entitled "Stories of Infinity."

14. Ramsaye, *Million and One Nights*, 162.

15. In an unpublished paper, "'A kind of shock to the retina': H. G. Wells, the Cinematograph, Automatic Electro-Photography and the Nervous Body" (2006). See also Shail's edited collection *Reading the Cinematograph: The Cinema in British Short Fiction, 1896-1912* (Exeter: University of Exeter Press, 2011) and monograph *The Cinema and the Origins of Literary Modernism* (New York: Routledge, 2012). Wells's prescience about cinema is noted by Keith Reader, in *H. G. Wells, Modernity and the Movies* (Liverpool: Liverpool University Press, 2007), and discussed by Laura Marcus in "The Coming of Cinema," in *Late Victorian into Modern*, ed. Laura Marcus, Michele Mendelssohn, and Kirsten E. Shepherd-Barr (Oxford University Press, 2016), 573-75.

16. H. G. Wells, "A Story of the Days to Come," in *The Short Stories of H. G. Wells* (London: Ernest Benn, 1927), 758.

17. In many ways *The Sleeper* was Wells's apocalyptic ur-text, which he revised in 1911 and partially recycled in later works, including *The Shape of Things to Come*.

18. See pp. 95-96 later in this chapter.

19. H. G. Wells, *When the Sleeper Wakes* (New York and London: Harper & Brothers, 1899), 150. This original text was revised and reissued in 1910 as *The Sleeper Awakes*, a version that has largely superseded the original. References here are all to the 1899 text, available in Leon Stover, ed., *When the Sleeper Wakes. A Critical Text of the 1899 New York and London First Edition* (Jefferson, NC: McFarland & Company, 2000).

20. Wells, *Sleeper Wakes*, 167.

21. Wells, *Sleeper Wakes*, 224.

22. Edison, letter to *Century Magazine*, June 1894. Later as a foreword to W. K. L. Dickson and Antonia Dickson, *History of the Kinetograph, Kinetoscope, and Kineto-phonograph* (New York: Albert Burn, 1895).

23. On the cultural impact of these fictions, see my "Early Phonograph Culture and Moving Pictures," in *The Sounds of Early Cinema*, ed. Richard Abel and Rick Altman (Bloomington: Indiana University Press, 2001).

24. In, respectively, *20,000 Leagues Under the Sea* (1870) and *Robur-le-Conquérant* (1886), known in English as *Robur the Conqueror* or, more commonly, *The Clipper of the Clouds*.

25. On world's fairs as spatiotemporal display, see Christie, *The Last Machine* (BBC TV, 1994), episode 2. Brown and Anthony cite a number of "multi-sensory presentation" patents granted in the 1890s as evidence to challenge Paul's "originality." Brown and Anthony, *The Kinetoscope: A British History* (East Barnet: John Libbey, 2017), 89–91. Undoubtedly such presentations and experiences were a popular aspiration of the fin-de-siecle, but only Paul seems to have taken inspiration from Wells's "journey" through cosmic time.

26. See appendix A.

27. On panoramas and panoramic painting, see Christie, "Kings of the Vast," *Tate Etc.*, Autumn 2011.

28. Moving the spectators' seats is retained in the present-day Atlanta Panorama.

29. The Crystal Palace Gardens included the Italian terraces, with massive dinosaur sculptures that were commissioned to accompany the relocated Great Exhibition that opened in 1854.

30. See Christie, "Contextualising Paul's 'Time Machine,'" *Cinema & Cie*, no. 3 (2003), 49–57.

31. Conventionally, the pamphlet by Boleslas [Bolesław] Matuszewski, *Une nouvelle source de l'histoire* (Paris, 1898), is often cited as the first claim for film as historical record. However, Paul's correspondence with the British Museum predates this by two years.

32. This episode was first revealed by Stephen Bottomore in "'The Collection of Rubbish': Animatographs, Archives and Arguments: London, 1896–97," *Film History* 7 (1995): 291–96.

33. Almost all films until the 1950s were printed on cellulose nitrate, which is unstable and combustible—hence the "nitrate problem" that haunted film archives for decades. But the custom of storing its successor, cellulose acetate, in metal cans was found to lead to a reaction known as "vinegar syndrome," which attacked the integrity of the stored film. Storage in glass containers would indeed be an ideal system.

34. *The Era*, 17 October 1896, 19; quoted in Bottomore, "Collection of Rubbish."

35. *Sea Cave near Lisbon* is the earliest work in the British copyright collection of films at Stationer's Hall, March 1897. See Richard Brown, "The British Film Copyright Collection," *Journal of Film Preservation*, no. 54 (1997), 52.

36. See Christie, "'What Is a Picture?' Film as Defined in British Law before 1910," in *Beyond the Screen: Institutions, Networks and Publics of Early Cinema*, ed. Marta Braun et al. (New Barnet: John Libbey, 2012), 78–84. Also chap. 9, pp. 192–96.

37. Cecil Hepworth, *Amateur Photographer* 25, no. 657 (1897): 374; quoted in John Barnes, *The Beginnings of the Cinema in England* (1976; Exeter: University of Exeter Press, 1998), 1:227.

306 NOTES TO PAGES 96-105

38. Barnes, *Cinema in England*, 2:10.

39. William Stanley, *Oxford Dictionary of National Biography*. See also https://www.yourlocalguardian.co.uk/news/4275330.croydon-legend-being-erased-from-history-books/.

40. Ramsaye was perhaps the first to claim that the Bazar de la Charité fire had a prejudicial effect on attitudes to film, stating that "it became . . . unfashionable to patronize the films," apparently based on "a letter written by an American observer" in Paris (*Million and One Nights*, 357). But he cites no other evidence, and apart from increased attention to fire safety, it would seem that attendance at most kinds of film show continued to increase.

41. *Le Figaro*, 5 May 1897.

42. Figures given for the death toll from the bazaar fire vary because some died after the actual fire and some were unidentifiable. The total seems to have been about 132.

43. Sévérine, in *L'echo de Paris*, 14 May 1897.

44. *Times*, 12 May 1897.

45. *Times*, 12 May 1897, 7.

46. *Optical and Magic Lantern Journal* 8, no. 99 (August 1897); quoted in Barnes, *Cinema in England*, 2:16.

47. *Amateur Photographer*, 3 September 97; Barnes, *Cinema in England*, 2:19.

48. In 1936, Paul recalled that "in [the] year before the Jubilee, public interest in animated pictures seemed to be on the wane." Paul, "Before 1910," *Proceedings of the British Kinematograph Society*, no. 38 (1936), 5.

49. Paul, "Before 1910."

50. G. W. Steevens, "Up They Came," *Daily Mail*, 23 June 1897. The fullest account of the Jubilee and its filming is in Luke McKernan, "Queen Victoria's Diamond Jubilee" (2012), http://lukemckernan.com/wp-content/uploads/queen_victoria_diamond_jubilee.pdf.

51. See Thomas Rooke, *Burne-Jones Talking: His Conversations 1895–98* (London: John Murray, 1982).

52. William Slade, the Cheltenham-based traveling exhibitor, made Jubilee films (although not Paul's) his main attraction during 1897–1898. See Patricia Cook, "Slade's Electro-Photo Marvel: Touring Film Exhibition in Late Victorian Britain," PhD thesis, Birkbeck College, 2016.

53. See chap. 7.

54. Society of Arts meeting, reported in the *Electrical Review*, 29 June 1901, 820.

Chapter 5

1. According to an undated and unsigned British Film Institute document, Mrs. Paul played the wife in *Come Along, Do!* (1898), was shown "tackling a burglar in *His Brave Defender*" (1902), and was "featured" in *The Lover and the Madman* (1905). Microfiche, BFI Library.

2. A handwritten letter from the Pauls to Irene Codd's mother, dated April 1938, offers condolences on the death of her husband, Arthur.

3. This historic theater had been renovated by Henry John Nye Chart in 1866, and after his death in 1876, his wife took over its management, becoming the first female theater owner in Britain.

4. Initially known simply as "Tub," this delightful comedy obviously had enough appeal to appear under its full title in Paul's large 1901 catalogue.

5. The marriage certificate records that their marriage took place in the St. Giles Register Office on 3 August 1897.

6. Frascati's, opened at 32 Oxford Street in 1893, had a high reputation for both its cuisine and its elaborate décor. For an account of dining there in 1899, see http://www.victorianlondon.org/publications2/dinners-30.htm.

7. Irene Codd papers, British Film Institute, Special Collections, 130. This would correspond with Paul establishing a New York branch of the Instrument Company ca. 1913.

8. W. H. Eccles, an electronics expert and friend of Paul's, in his obituary, "Robert W. Paul: Pioneer Instrument Maker and Cinematographer," *Electronic Engineering*, August 1943.

9. The Women Film Pioneers Project has been working to correct this misperception for some years. See https://wfpp.cdrs.columbia.edu/ and the recent book by one of its founders, Jane Gaines, *Pink-Slipped: What Happened to Women in the Silent Film Industries?* (Urbana: University of Illinois Press, 2018).

10. Patricia Cook, "Slade's Electro-Photo Marvel: Touring Film Exhibition in Late Victorian Britain," PhD thesis, Birkbeck College, 2016.

11. Paul created a new access entrance to his expanding premises in Sydney Road, which he named Newton Road in 1901. "Application for photographic laboratories R Paul, submitted by Victor C Jackson, 4 Finsbury Square"; London Metropolitan Archives, LMA/4070/02/00695.

12. This was part of what might be termed "the oral tradition" in the 1990s, and although not recorded in any text, led to early records searches for details of what was then believed to be an only child. See my "A Robert Paul Newsletter," January 2004, recording the then state of family information.

13. These films are not known to survive, although stills appeared in Paul's 1898 catalogue. See chap. 8, p. 000.

14. See chap. 8.

15. Fiction production resumed in September 1899, much of it using "trick" effects, which may well have led to hiring the conjuror Walter Booth.

16. Lizzie Paul was born on 26 February 1902, but died within an hour, due to "asphyxia owing to umbilical cord round the neck during labour." George Rollason Paul was born prematurely on 17 July 1903, but only survived three days, with the cause of death given as "asthenia."

17. Codd papers (BFI), 130. Paul's brothers predeceased him, in 1919 and 1922, and his will included bequests to two nephews.

18. George Ivens Paul's address is given in 1943 as Tarwo, The Gold Coast; and David Lyon Paul's as 74 Edgbaston Road, Birmingham.

19. George Ivens Paul was born to Arthur and Beatrice Paul, and baptised in Saltford, Somerset, on 10 June 1911.

20. *Optical Lantern and Cinematograph Journal*, July 1905.

21. Eccles, "Robert W. Paul."

22. Codd papers (BFI), 147. On the cafe's colorful history, see Guy Deghy and Keith Waterhouse, *Café Royal: Ninety Years of Bohemia* (London: Hutchinson, 1955), and on its closure, Stephen Bates, "Cafe Royal Party Is Over as 143 Years of High Society Goes under the Hammer," *Guardian*, 23 December 2008.

23. "Animatograph Jollies," *Optical Lantern and Cinetograph Journal*, August 1906, 186. I am especially grateful to Simon Popple for this reference.

24. These included Howard Cricks, J. H. Martin, Jack Smith, and Frank Mottershaw.

25. Codd papers (BFI), 44.

26. Codd papers (BFI), 45–46.

27. *The Fatal Hand* (1907) was identified in the archival collection of the Swedish Film Institute in 2016 and restored under the supervision of Camille Blot-Wellens.

28. This may have been foreseen earlier, as plans for building a new "detached house" were submitted to Friern Barnet Urban District council in 1908.

29. Codd papers (BFI), 145.

30. See chap. 6, pp. 132–35.

31. G. Watchorn, *Never a Dull Moment: A History of the Cambridge Instrument Company 1918–1968* (Cambridge, 1990), 15.

32. Codd papers (BFI), 145.

33. Codd writes that he refused the knighthood, saying "that his family was old enough for him to remain plain Mr Paul." Codd papers (BFI), 138.

34. See chap. 11.

35. Paul, "Before 1910," *Proceedings of the British Kinematograph Society*, no. 38 (1936).

36. For details of Winnington Ingram's campaigns against sexual immorality, particularly identifying cinemas as places for homosexual encounters, see Dean Rapp, "Sex in the Cinema: War, Moral Panic and the British Film Industry, 1906–1918," *Albion*, no. 53 (Fall 2002), 436.

Chapter 6

1. Robert Paul, "Before 1910," *Proceedings of the British Kinematograph Society*, no. 38 (1936), 4.

2. "Before 1910," 4.

3. Paul's program at the Alhambra on 31 August 1896 has survived as a useful example of what was presumably the standard music-hall film offer at this time. Assuming the films averaged 40 feet, this would amount to 800 feet total, equivalent to 12 minutes at 16 frames per second (although probably closer to 18 minutes with time required for threading each new film). By 1907 films were longer, so fewer were needed. See Theodore Brown's sample programs in *Kine Weekly*, 22 August 1907, 229.

4. For a typical British film program of 1909, see Ian Christie and John Sedgwick, "'Fumbling towards Some New Form of Art?' The Changing Composition of Film Programmes in Britain, 1908–1914," in *Film 1900: Technology, Perception, Culture*, ed. Annemone Ligensa and Klaus Kreimeier (New Barnet: John Libbey, 2009), 154.

5. For details of Smith's performances and career, see Erkki Huhtamo, *Illusions in Motion: Media Archaeology of the Moving Panorama and Related Spectacles* (Cambridge, MA: MIT Press, 2013), 215–44.

6. On Holmes and lantern travelogues, see Theodore X. Barber, "The Roots of Travel Cinema: John L. Stoddard, E. Burton Holmes and the Nineteenth-Century Illustrated Travel Lecture," *Film History* 5, no. 1 (March 1993): 68–84.

7. Lumière subjects were intrinsically topographical, offering "views" (*vues*) of places and activities, and were sometimes presented in thematic groups, such as an

eight-part bullfight series (*Course de taureaux*) at the Grand Café in Paris in December 1898. But more often, and certainly in 1896–1897, Lumière programs stressed variety, as in Montpellier on 30 August 1896. Jacques André and Marie André, *Une saison Lumière a Montpellier* (Perpignan: Institut Jean Vigo, 1987), 126.

8. Charles Musser and Carol Nelson, *High Class Moving Pictures: Lyman Howe and the Forgotten Era of Travelling Exhibition, 1880–1920* (Princeton, NJ: Princeton University Press, 1991); Vanessa Toulmin, "The Importance of the Programme in Early Film Presentation," *KINtop*, no. 11, "Kinematographen-Programme" (Autumn 2002): 19–34. See also Christie, "On Programmes," Lumière centenary conference, Université Lyon 2, 1995.

9. My own research on Hertz's itinerary and presentations has benefited from that of Chris Long, in "Australia's First Films," in *Cinema Papers* (1993), and of Tony Martin-Jones, on his "Carl Hertz in Australia" website, http://www.apex.net.au /~tmj/carl-hertz.htm.

10. A version of this, from the *Referee* in Sydney, 7 October 1896, is quoted by Martin-Jones.

11. At a projection speed of 16 frames per second, a minute-long film would have had just under a thousand frames. Paul's price per film at this time was between £3 and £6, so a program of ten could have cost an average of £45.

12. For recent research on the lively competition in Australia in 1896, see Sally Jackson, "Like Boils the Cinématographe Tends to Break Out: The Films of the 1896 Melbourne Cup: The Lawn Near the Bandstand," *Screening the Past* (2015), http://www .screeningthepast.com/2015/06/like-boils-the-cinematographe-tends-to-break-out -the-films-of-the-1896-melbourne-cup-the-lawn-near-the-bandstand/.

13. Graham Shirley and Brian Adams, *Australian Cinema: The First Eighty Years* (Sydney: Angus and Robertson, 1983), 7.

14. Clive Sowry, "Film Pioneers of New Zealand, no 9: The Kinematograph Arrives in New Zealand," *Big Picture*, no. 6 (1996). However, as far as is known, the first commercial film screening in New Zealand took place on 7 November 1896, when J. F. Macmahon showed moving pictures on "Edison's Cinematographe" in High Street, Christchurch. The program included *Traffic in Broadway, Wheelwright at Work*, and *Sandow the Strong Man*.

15. Until recently, information about Rousby has been scant and often confused. But research by Luis Guadano has revealed that he was indeed both Hungarian-born (birth name Sandor) and American. Luis Guadano, "Edwin Rousby: A Mystery Unveiled," in *Filmhistoria Online* 24, no. 2 (2014).

16. Guadano, "Mystery Unveiled."

17. *La Correspondencia de España*, 18 May 1896, quoted in Guadano, "Mystery Unveiled."

18. Seguin and Letamendi, *Los orígenes del cine en Cataluña* (Barcelona: Generalitat de Catalunya, 2004), 68; quoted in Guadano, "Mystery Unveiled."

19. "The Atlantic coast is the nexus between the United Kingdom and the Spanish interior." On the transportation logic of Short's route, see Begonia Soto, "Rethinking Boundaries: The First Moving Images between Spain and Portugal," in *Networks of Entertainment: Early Film Distribution 1895–1915*, ed. Frank Kessler and Nanna Verhoeff (Eastleigh: John Libbey, 2007), 11–12.

20. Code-named "Cave" in the 1901 catalogue, which also included *A Storm in Dover Harbour*.

21. *New York Herald*, 24 April 1896, 11. Quoted by Charles Musser, *Before the Nickelodeon: Edwin Porter and the Edison Manufacturing Company* (Berkeley: University of California Press, 1991), 63. Paul had sent a copy of this film, produced during his partnership with Acres, to Edison. The other five films in the Broadway show were all taken inside the Black Maria studio.

22. Georges Sadoul, *Lumière et Méliès* (Paris: L'herminier, 1985), 30.

23. Stasov, writing on 30 May 1896, quoted in Jay Leyda, *Kino: A History of the Russian and Soviet Film* (London: George Allen & Unwin, 1960), 18.

24. A Filoscope version, held by the Bill Douglas Collection at Exeter University, was digitally animated for the British Film Institute DVD of Paul's *Collected Films* in 2008.

25. I owe the discovery of this to Stephen Bottomore, who wrote about it in "George R. Sims and the Film as Evidence," in *Reading the Cinematograph: The Cinema in British Short Fiction, 1896–1912*, ed. Andrew Shail (Exeter: University of Exeter Press, 2011).

26. Bottomore, "Film as Evidence," 30–31.

27. Only two of these survive, *Women Fetching Water from the Nile* and *Fisherman and Boat at Port Said*, but the other titles hardly promised either exotic or spectacular scenes, with two exceptions noted below (see filmography, 1897).

28. Hepworth, writing in the *Amateur Photographer* 25, no. 658 (14 May 1897). In 1908, however, Hepworth would release a series of Egyptian subjects.

29. So rare are early images of this ritual that a postcard collection has been proposed for UNESCO World Heritage adoption; http://english.ahram.org.eg/News Content/32/138/232771/Folk/Photo-Heritage/Egypt--years-ago-The-Ceremony -of-the-Holy-Carpet-M.aspx. The earliest surviving film appears to be from 1918; http://www.colonialfilm.org.uk/node/6160.

30. Opera House advertisement, *Melbourne Herald*, 14 August 1897.

31. *Sun* (Melbourne), 27 August 1897, 5.

32. See chap. 8. Also Christie, "'As in England': The Imperial Dialogue in Early Film," in *Domitor 2008: Peripheral Early Cinema*, ed. François-Amy de la Breteque et al. (Perpignan: Presses Universitaires de Perpignan, 2010), 223–29.

33. "Before 1910," 4. In this paper, the visit to Sweden is included in a section headed "Selling projectors in 1896," but from the date of the films being advertised, it clearly happened in the summer of 1897.

34. See Anders Ekstrom, Solveig Jülich, and Pelle Snickars, eds., *1897 Mediehistorier kring Stockholmsutställningen* (Stockholm: Statens BilmArchiv, 2005), esp. Snickars, "Mediearkeologi" (Media Archaeology), 125–63. Also Allan Pred, *Recognising European Modernities: A Montage of the Present* (London: Taylor and Francis, 1995).

35. Cheltenham Cricket Club playbill, reproduced in Rachael Low and Roger Manvell, *The History of the British Film*, vol. 1, *1896–1906* (London: Allen & Unwin, 1948), plate 11.

36. Toulmin, "Importance of the Programme," 26–27.

37. Released in August 1898, the Constantinople series comprised *Embarking*, *The Scene of Armenian Massacres*, and *A Street in Stamboul*. The massacre referred to would have been the pogrom of 1894–1896, which was a prelude to the Armenian genocide of 1915.

38. Barnes gives details of Prestwich's films in *The Beginnings of the Cinema in En-*

gland (1976; Exeter: University of Exeter Press, 1998), 3:25–29. Only Prestwich's and Paul's films survive, so it is not clear if Wolff had original film or was distributing the work of others.

39. The coaling film was code-named "Cory," making clear that it showed how Cory & Sons had added to the "batteries of hydraulic cranes on the foreshore" with a first floating derrick in 1860. See Michael J. Rustin, *London's Turning: The Making of Thames Gateway* (London: Routledge, 2016), 22.

40. *Daily Chronicle*, 23 June 1898.

41. *Daily Chronicle*, 24 June 1898.

42. Reported in the *British Journal of Photography* 45, no. 1991 (1 July 1898): 449.

43. *Amateur Photographer* 27, no 717 (1 July 1898).

44. *Daily Chronicle*, 25 June 1898; *The Era*, 25 June. See also Barnes, *Cinema in England*, 3:20–22, 182.

45. Paul letter, 25 June 1898.

46. *Photogram* 5, no. 56 (August 1998).

47. See, for instance, André Bazin's famous and influential essay "The Ontology of the Photographic Image" (1945), which links photography to the psychology of relics and memorialization. In Hugh Gray, ed., *What Is Cinema?* (Berkeley: University of California Press, 1967), 9–22.

48. "When this apparatus is on sale to the public . . . death will cease to be absolute" (*La poste*, 30 December 1895). "It will be possible to see our nearest alive again long after they have gone" (*Le radical*, 30 December 1895). Quoted in Christie, *The Last Machine: Early Cinema and the Birth of the Modern World* (London: BBC Educational Projects and BFI Publishing, 1994), 111; Emmanuelle Toulet, *Cinématographe, invention du siècle* (Paris: Decouvertes Gallimard, 1988), 134–34.

49. See, for instance, Kipling's "Mrs Bathurst" (1904), in which a sailor is driven mad by the on-screen image of a woman he believes is looking for him, or Apollinaire's "Un beau film" (1907), in which a syndicate stage sensational murders on camera.

50. See chap. 8.

51. Hale's Tours made its debut at the 1904 St. Louis exposition (where, no doubt coincidentally, Paul's electrical instruments won an award), and by 1906 had installations in many countries around the world. In London, it was established at 165 Oxford Street. By 1908, however, its novelty had worn off, and the company went into receivership, with only the London venue remaining open until 1910. See entry in *The London Project*. AHRB Centre for British Film and Television Studies, http://londonfilm .bbk.ac.uk/view/business/?id=394. Also http://www.londonssilentcinemas.com /westendexhibts/hales-tours/.

52. *Process Engraver's Monthly* 8 (1901): 254, cited and illustrated in Roland-Francois Lack's blog, http://www.thecinetourist.net/robert-paul-in-london-tour -guide-and-film-maker.html.

53. *Photographic Times*, 1903.

54. Lack blog, https://www.thecinetourist.net/robert-paul-in-london-tour-guide -and-film-maker.html.

55. The series of short travelogues made by Harry B. Parkinson and Frank Miller in 1924 has now been restored by the BFI National Archive. Titles include *Barging through London, Cosmopolitan London, London's Sunday, Flowers of London, London's Free Shows,* and *London off the Track.*

56. Charles Urban, an American resident in Britain from August 1897, launched

his Charles Urban Trading Company in 1903 and quickly became the leading distributor of nonfiction films, publishing an impressive annual catalogue. See Luke McKernan, *Charles Urban: Pioneering the Non-Fiction Film in Britain and America, 1897–1925* (Exeter: University of Exeter Press, 2013).

57. For example, *A Collier's Life* (1904), *The Life of a Racehorse* (1904), *Life of a Railway Engine* (1904), and even *The Adventures of a £100 Banknote* (1905).

58. The Peek Frean film runs for 36 minutes. See http://www.screenonline.org.uk /film/id/711535/index.html.

59. *Optical Lantern and Cinematograph Journal*, October 1905, 260.

60. *Optical Lantern and Cinematograph Journal*, September 1906.

61. *Aberdeen Press and Journal*, 2 October 1906.

62. Other Irish titles, all released between September and December 1908, included *The Falls of Doonas*; *Salmon Fishing*; *Village Life*; *Lobster Fishing*, apparently showing coracle fishermen off northwest Ireland; and *A Cattle Drive in County Galway*, offering a reconstruction of cattle rustling.

63. *Kine Weekly*, 11 July 1907, 1, 9.

64. *Kine Weekly*, 22 August 1907, 229.

65. The nonfiction films in this case, *Shark Fishing in the North Sea* and *Trollhatten Falls in Winter*, were of the type that Paul had been producing, but such subjects were destined to play a decreasing part in programming. See Christie and Sedgwick, "Fumbling towards Some New Form of Art?"

66. *Kine Weekly*, 10 September 1908, 403.

67. Smith was later reported to have left Warwick to rejoin Paul, and to have re-edited a film commissioned by the *Football Star*, released by Paul as *How Football Reports Are Prepared* early in 1909, around the same time as his *Morning Leader* film. Pathé is usually credited with having pioneered the newsreel in its *Journal*, launched in 1911, although a venue called the Daily Bioscope had opened in London in 1906, showing mainly topicals. See Jon Burrows, "Penny Pleasures: Film Exhibition in London during the Nickelodeon Era, 1906–1914," *Film History* 16, no. 1 (2004): 78.

68. *Kine Weekly*, 8 October 1908, 501.

Chapter 7

1. *Times, Daily Telegraph* editorials, 11 October 1899.

2. The first Boer War resulted from Britain annexing the Transvaal Boer republic in 1877, which provoked Paul Kruger to rebel and inflict a heavy defeat on British forces at Majuba Hill in February 1881.

3. *London Daily News*, 12 October 1899, 4.

4. *Arrival of the Sirdar at Dover* and *Reception of the Sirdar at Guildhall*. "Sirdar" was the title given to the commander of the British-supervised Egyptian Army.

5. The term "jingoism" came from a music-hall song of the 1877–1878 Russo-Turkish War, "We don't want to fight, but by jingo if we do, we've got the ships" etc.

6. Thomas Pakenham, *The Boer War* (1979; Abacus, 1991), 283.

7. Barnes, *Filming the Boer War*, vol. 4 of *The Beginnings of the Cinema in England* (1976; Exeter: University of Exeter Press, 1998), 7.

8. *Windsor Magazine*,1900, cited in Simon Popple, "'But the Khaki-Coloured Camera Is the Latest Thing': The Boer War Cinema and Visual Culture in Britain," in *Young*

and Innocent? The Cinema in Britain 1896–1930, ed. Andrew Higson (Exeter: University of Exeter Press, 2002), 13–27.

9. Stereographic photographs had been a major Victorian medium since the 1860s, and one of the largest companies in the business, Underwood & Underwood, published a set of Boer War stereographs. See H. J. Erasmus, *The Underwood & Underwood Stereographs of the Anglo-Boer War, 1899–1902*, http://scholar.sun.ac.za/handle /10019.1/82250?show=full.

10. Chamberlain to the Chancellor of the Exchequer, 7 October 1899; Pakenham, *Boer War*, 84.

11. Pakenham, *Boer War*, 100.

12. Diary of Edward Cutler, Scots Guards, http://www.angloboerwar.com/unit -information/imperial-units/659-scots-guards?showall=&start=2.

13. *The Era*, 14 October 1899, 27.

14. Published in the *Daily Mail* and elsewhere. Kipling raised money for supporting soldiers by waiving his copyright.

15. Richard Brown and Barry Anthony, *A Victorian Film Enterprise. The History of the British Mutoscope and Biograph Company, 1897–1915* (Trowbridge: Flicks Books, 1999), 48 ff.

16. On this important genre, see Ian Christie, "'The Captains and the Kings Depart': Imperial Departure and Arrival in Early Cinema," in *Empire and Film*, ed. Lee Grieveson and Colin Mccabe (London: Palgrave/British Film Institute, 2011).

17. Paul, *The Hundred Best Animated Photograph Films*, season 1900–1901, 11.

18. According to the *Hornsey Journal* (London), 9 December 1899. See Christie, "The Anglo-Boer War in North London," in *Picture Perfect; Landscape, Place and Travel in British Cinema before 1930*, ed. Laraine Porter and Bryony Dixon (Exeter: University of Exeter Press, 2007), 82–91.

19. See, for instance, nos. 5469–71 of the Warwick listing given by Barnes, *Cinema in England*, 4:283.

20. On the Muswell Hill lecture and other manifestations of the war in North London, see Christie, "Anglo-Boer War in North London."

21. Reviewed in the *Yorkshire Post*, 20 March 1900; quoted in Popple, "Khaki-Coloured Camera," 25.

22. The Tees' poster—dated "Tuesday Feb. 27 1899," clearly due to a printer's failure to set 1900—advertises a mixture of Cape actualities and "reproduction" scenes, together with "A Selection of Dissolving Views." Barnes, *Cinema in England*, 4:82.

23. *Hornsey Journal*, 16 December 1899.

24. As in the Muswell Hill show, most of the films appear to come from the Warwick catalogue, although the first two or three may have been from Fuerst, and *Coaling by Natives* could be a Biograph subject, if these were already on sale to other exhibitors.

25. Editorial, *Hornsey Journal*, 23 December 1899.

26. Eric Hobsbawm, *Age of Empire, 1875–1914* (London: Weidenfeld, 1987), 161.

27. Paul's brothers, Arthur and George, were twenty-six and twenty-two, respectively, when they volunteered, and both survived the Boer War, although both died relatively young (Arthur in 1922, George in 1919).

28. In a letter of 20 September 1937 to Thelma Gutsche, kindly made available by John Barnes. Gutsche would use this in her *The History and Social Significance of*

Motion Pictures in South Africa 1895–1940 (Cape Town: H. Timmins, 1972). See also Paul, "Before 1910," *Proceedings of the British Kinematograph Society*, no. 38 (1936), 5.

29. Paul, *Hundred Best Animated Photography Films*, 12.

30. During 1901, Emily Hobhouse visited and reported on the "sixty thousand men, woman and children stuffed into 'refugee' camps set up by Kitchener." Pakenham, *Boer War*, 502.

31. From Paul's 1901 catalogue.

32. Paul admitted in 1936 that he could not "vouch for the description of [his reproductions] by the showmen." Paul, "Before 1910," 5.

33. The first five advertised in *The Era* on 25 November 1899, and the remainder on 9 December.

34. Thanks to the New Zealand Film Archive, it was possible to include *A Camp Smithy*, the first of Paul's "reproductions" to be identified, in the BFI DVD *Collected Films* (2007). Clive Sowry listed this among "a very special collection" of films given to the New Zealand National Film Unit by the Auckland Museum of Transport and Technology in 1980. *Archifacts*, Bulletin of the Archives and Records Association of New Zealand, no. 16, December 1980.

35. See for instance, *Boers Bringing in British Prisoners* and *Capture of Boer Battery by the British* (both Edison, 1900) and Mitchell & Kenyon's *A Sneaky Boer* (1901).

36. Quoted from the *Yorkshire Post* in Popple, "Khaki-Covered Camera," 24.

37. Advertisement for Montgomery's Kinematograph and Concert Company, Auckland Opera House, *Evening Post* 60, no. 16 (19 July 1900): 6, http://paperspast .natlib.govt.nz/cgi-bin/paperspast?a=d&d=EP19000719.2.60.3&dliv=&e=.

38. "A Cinematographe of the War," *Argus*, 12 February 1900.

39. All of these are lost and known only from catalogue illustrations and extended descriptions.

40. See Frank Gray, "The Vision Scene: Revelation and Remediation," in *The Image in Early Cinema: Form and Material*, ed. Scott Curtis, Philippe Gauthier, Tom Gunning, and Joshua Yumibe (Bloomington: Indiana University Press, 2018), 36–45.

41. What Jay David Bolter and Richard Grusin call "remediation." Bolter and Grusin, *Remediation: Understanding New Media* (Cambridge, MA: MIT Press, 1999).

42. Britannia, derived from the Roman name for the British Isles, was revived as an allegorical personification of the United Kingdom after the Scottish king James became James I of England, Scotland, and Wales in 1603. "Rule Britannia," originally part of the masque *Alfred* (libretto by James Thompson, music by Thomas Arne, 1740), became an important patriotic song during the Georgian and Victorian period, celebrating British naval power. The image of a helmeted Britannia has appeared continuously on British coins since the late eighteenth century.

43. *Britannia Triumphans*, devised in 1638 by Sir John Davenant for Charles I, with designs by Inigo Jones, was one of the most lavish of court masques, its imagery linked with Rubens's decoration of the Banqueting Hall at Whitehall Palace. See Roy Strong, *Britannia Triumphans: Inigo Jones, Rubens and the Whitehall Palace* (London: Thames and Hudson, 1981).

44. Perhaps significantly, the first challenge to a long-standing monopoly of licensed "patent" theaters in London was the Britannia, opened in 1841 by Sam Lane, which became a bastion of working-class theater. More recently, a Britannia sketch had been presented at the Oxford music hall in 1885, as described by Penny Summer-

field, "Patriotism and Empire: Music Hall Entertainment 1870–1914," in *Imperialism and Popular Culture*, ed. John MacKenzie (Manchester: Manchester University Press, 1986), 28–29.

45. For an overview of the evolution of the Britannia allegory, see Marina Warner, *Monuments and Maidens: The Allegory of the Female Form* (London: Weidenfeld and Nicolson, 1985), 45–49.

46. Paul had already filmed a performance by Charles Bertram, the court conjuror, entitled *Britannia: Hail Britannia!*, in 1898. See his *Illustrated Catalogue of Animated Photograph Films* (August 1898), no. 15, quoted by Barnes, *Cinema in England*, 3:176.

47. Paul, *Hundred Best Films*, 1.

48. A music-hall example from the Crimean War is the "Inkerman *scena*," described by Dave Russell in "'We Carved Our Way to Glory': The British Soldier in Music Hall Song and Sketch, 1880–1914," in *Popular Imperialism and the Military*, ed. John MacKenzie (Manchester: Manchester University Press, 1992), 69–71.

49. The film is actually billed in Paul's 1901 catalogue as a "Patriotic Song with Animated Illustrations," with words by Clarence Hunt and music by Frank Byng, one of four such offerings.

50. According to the *Daily News* review, the Alhambra screening on 18 September ran for "an hour and a quarter."

51. From a review in *Today*, quoted in Paul's brochure.

52. From Paul's 1901 catalogue (British Film Institute). Only one part of *Army Life* is known to exist at present: a gallop by *Mounted Infantry*.

53. Richard Brown, "War on the Home Front: The Anglo-Boer War and the Growth of Rental in Britain. An Economic Perspective," *Film History* 16, no. 1, "Early British Cinema" (2004): 28–36.

54. Williamson's film may not have originated in this form, according to a detailed analysis in Martin Sopocy, *James Williamson: Studies and Documents of a Pioneer of the Film Narrative* (London: Associated University Presses, 1998), 39–45.

55. T. C. Hepworth, "Music and 'Effects' in Cinematography," *Showman*, 6 September 1901.

56. After killing their captain, most of the crew of the *Aurora* joined the Bolsheviks, and the ship fired a blank to signal the start of the assault on the Winter Palace. Thereafter the ship became iconically linked with the October Revolution.

57. The loss of life and reaction in Hull attracted wide attention, with the local newspaper reporting that funeral spectators "felt they were taking part in an event which the whole country was viewing." "A City's Grief," *Eastern Morning News*, 28 October 1904, 5.

58. This was noted by Georges Sadoul in his *Histoire générale du cinema*, vols. 1–2 (Paris: Denoël, 1946–1948). See chap. 11.

Chapter 8

1. Robert Paul, "Before 1910," *Proceedings of the British Kinematograph Society*, no. 38 (1936), 5.

2. Gifford lists 101 "fictional" titles published in 1898, of which Paul produced 23, with G. A. Smith of Brighton the next most prolific producer, at 13, and others supplying ten or less. However, in this period Gifford's distinction between "nonfiction" and "fiction" is somewhat arbitrary. His two volumes list a total of 660 titles for 1898,

among which I have found a total of 83 Paul productions, making him still the majority producer. Gifford, *British Film Catalogue*, vols. 1 and 2.

3. The Pathé studio and factory would eventually become a major employer in Chatou.

4. Advertisement in *The Era*, 8 October 1898, 27. Text published in John Barnes, *Cinema of England*, vol. 3, subtitled *1898: The Rise of the Photoplay* (1998), 16–17. Discussed by Roland-François Lack in his chapter in *London on Film*, ed. Pam Hirsch and Chris O'Rourke (Cham, Switzerland: Palgrave/Macmillan, 2017).

5. In late 1896, Méliès's *After the Ball* and *The Magician* both showed nearly naked women, while Alice Guy's *The Cabbage Fairy*, for Gaumont, had a woman producing babies from under cabbages.

6. We do not know in what order, or over what period, the twenty films listed by Gifford as released in August 1898 were made. It is possible that *The Sailor's Return* was made last, before the 23 August tragedy, which was followed by a pause in Paul's output of over a year, until the following September.

7. Only the first shot of this film has survived, but two frames of the second appear in Paul's catalogue (and were added to the version included on the BFI DVD).

8. David Robinson has traced a rich prehistory of versions of *Come Along, Do!* in a wide variety of media. An 1874 American song with this title by Charles Miers tells of taking "the Missus . . . to Woods Museum . . . to study the classical nude." The wife finds Venus and Jupiter "undressed and decidedly rude," insisting that the man turn away. See Library of Congress record, at https://www.loc.gov/resource/sm1874.04741.0/?sp=3.

9. The claim that the Pauls appeared in the film is recorded on a British Film Institute microfiche, with no source given. Both figures do seem made up to appear older than they are.

10. In nonfiction, there were already examples of films that offered an elliptical view of an event, with "invisible" breaks in the continuity of shots, as in Acres's pioneering 1895 Derby film and a number of Lumière subjects.

11. In his program "The Magic Films of Robert W. Paul" for the Pordenone Giornate del Cinema Muto, 2016, William Barnes announced the visual identification of Walter Booth, based on two portraits in the John Salisse Archive, held by the Davenport Collection of material relating to the history of magic. Barnes was confident about identifying Booth in at least nine Paul films, adding "although it is likely that he appears heavily disguised in other roles also," having specialized in "quick change impersonations." See also chap. 11 on claims for Booth as director.

12. Codd refers to Paul having "painted backdrops for the following day's filming" after finishing his other work.

13. Browning's poem first appeared in his collection *Dramatis Personae* (1864) and was understood to be an attack on the well-known contemporary medium Daniel Home. See Isobel Armstrong, "Browning's Mr. Sludge, 'The Medium,'" *Victorian Poetry* 2, no. 1 (Winter 1964): 1–9. George and Weedon Gissing's comic novel *The Diary of a Nobody* first appeared in serial form in *Punch*, 1888–1889, then as a book in 1892. A chronicle of lower-middle-class life in London, it has become a classic, and includes scenes of domestic séances organized by Mrs. Pooter, of which her husband disapproves.

14. On the early history of the Egyptian Hall, in London's Piccadilly between 1812

and 1905, see primarily Richard Altick, *The Shows of London* (Cambridge, MA: Harvard University Press, 1978), 235–52.

15. Dickens had long formed part of the magic lantern repertoire, but the earliest moving-picture adaptation seems to have been W. K. L. Dickson's *The Death of Nancy Sykes*, from a vaudeville sketch based on *Oliver Twist*, filmed for Biograph before Dickson left the United States in April or May of 1897.

16. The extant version is incomplete, running for about one-third of the original ten minutes, including some original scene titles. Scene 2 shows Marley's face superimposed over Scrooge's door knocker, a vertical wipe-transition from bottom to top as Scrooge enters his house—probably the first ever seen—and "visions" of Scrooge's youth superimposed on a black curtain in his bedroom. This is recorded as a "milestone" at http://filmsite.org/visualeffects 1.html, with Walter R. Booth confidently identified as "director."

17. Vignetted inserts had first appeared in G. A. Smith's *Santa Claus* (1898), and stop-motion was used by Méliès and others since late 1896.

18. Jay David Bolter and Richard Grusin, *Remediation: Understanding New Media* (Cambridge, MA: MIT Press, 1999).

19. Pain's Pyrodrama at Alexandra Palace, London: *The Last Days of Pompeii* (1898). See also, on the history of the novel's adaptation on stage and in paintings, Maria Wyke, *Projecting the Past: Ancient Rome, Cinema and History* (London: Routledge, 1997).

20. From Paul's 1901 catalogue. See also my article, "*The Magic Sword*: Genealogy of an English Trick Film," *Film History* 16, no. 2 (2004): 163–71.

21. The fantastic ballets were a celebrated feature of the Alhambra, often featuring prominent dancers and with elaborate scenic and lighting effects. See, for instance, details of *The Sleeping Beauty*, as presented in 1891, at http://www.arthurlloyd.co.uk/Alhambra/1891Program.htm.

22. *The Enchanted Cup* (1902) and perhaps *A Knight Errant* (1907) seem to have been in a similar vein. For discussion of some of the process techniques in *The Magic Sword* and similar subjects, see Frederick A. Talbot, *Moving Pictures: How They Are Made and Worked* (London: Heinemann, 1912).

23. "Notes on Current Topics," *Kine Weekly*, 22 August 1907.

24. Paul, 1902 catalogue, BFI Library.

25. Stereoscopic illusion had been widely available throughout the later nineteenth century, and many early filmmakers hoped to create this effect in moving pictures—notably Louis Lumière. But when the problems of stereo projection and viewing appeared insuperable, it was realized that carefully composed moving images did create a quasi-stereoscopic effect, especially in diagonal compositions.

26. Jonathan Auerbach identifies two films from American Mutoscope and Biograph, *The Escaped Lunatic* (1903) and *Personal* (1904), as influential demonstrations of "the farcical potential of figures running frantically after one another." These would spark a flood of imitations, both in America and in France, where Pathé soon became a major purveyor of increasingly elaborate chase films. Auerbach, "Chase Films," in *Encyclopedia of Early Cinema*, ed. Richard Abel (Abingdon: Routledge, 2005): 110.

27. "New Southgate Motorist Fined," *Barnet Press*, 10 August 1907.

28. The *Evening Standard*, 26 July 1906, reported 1,245 fines for motorists. Conan Doyle's letter on police "ambushing" of motorists, "More Motorphobia," appeared in

the *Daily Mail*, 21 September 1905, https://www.arthur-conan-doyle.com/index.php ?title=More_Motorphobia. See also his letter on "Motor Cars and Coast Defence," *Times*, 12 April 1906; https://www.arthur-conan-doyle.com/index.php?title=Motor -Cars_and_Coast_Defence.

29. Reported in the *Barnet Press*, 10 August 1907, 6.

30. *Le Raid Paris–Monte Carlo* was commissioned from Méliès by the Folies-Bergeres for use in a stage extravaganza.

31. Henley's poem appeared as a pamphlet in 1903 (published by David Nutt, Long Acre, London). Kipling's fourteen parodic poems appeared in the *Daily Mail* in 1904.

32. "The Founding and Manifesto of Futurism," trans. R. W. Flint, in *Documents of 20th Century Art: Futurist Manifestos*, ed. Umbro Apollonio (New York: Viking, 1973), 19–24. Originally published in *Le Figaro* (Paris), 20 February 1909.

33. Andrei Bely, "The City," quoted in Yuri Tsivian, *Early Cinema in Russia and Its Reception* (London: Routledge, 1994), 150.

34. On Sims's 1897 story, see chap. 6, pp. 127–28. Rudyard Kipling's story "Mrs Bathurst" (1904) and Apollinaire's "Un beau film" (1907) have been extensively discussed as early cultural responses to film. Both these and other modernist responses to early film were explored in the television series and book *The Last Machine* (Christie, 1994).

35. According to IMDb, the leading part was played by Leah Marlborough, a daughter of the Punch and Judy pioneer Richard Codman, and the "earliest named actor" in a British film. However, Ellen Paul was also linked with the film (BFI microfiche).

36. Identified by Camille Blot-Wellens, in the Swedish Film Institute archive.

37. Gifford, and others, are confident that Martin was the main "director" for Paul during 1907–1908.

38. Paul has six films involving a tramp.

39. Chesterton's essay "A Defence of Detective Stories" appeared in 1901 and is discussed by John C. Tibbetts in "The Case of the Forgotten Detectives: The Unknown Crime Fiction of G. K. Chesterton," http://www.gkc.org.uk/gkc/detectives.txt.

40. *Kine Weekly*, 20 May 1909, 94.

41. Pioneered in *Fire!* (1901) by the Brighton filmmaker James Williamson, then extended in Edwin Porter's *The Life of an American Fireman* (1903) for Edison.

Chapter 9

1. The cartoon appeared in the *Optical Lantern and Cinematograph Journal*, September 1905, 236. I am indebted to Luke McKernan for discovering this cartoon.

2. Brown started and edited the *Kinematograph and Lantern Journal* in 1907.

3. Six other shows around London, according to Leslie Wood, *The Miracle of the Movies* (London: Burke Publishing, 1947), 101.

4. Edward. G. Turner, "From 1896 to 1926: Recollections of Thirty Years of Kinematography," *Kine Weekly*, 17 June 1926. The other manufacturer would probably have been Alfred Wrench, who patented his Cinematograph projector in August 1896, with a distinctive ratchet-and-pawl intermittent mechanism. This remained "the only serious rival to Paul's Theatrograph," in the opinion of Stephen Herbert. Alfred Wrench entry, *Who's Who of Victorian Cinema*, http://www.victorian-cinema.net/wrench.

5. "The Prince's Derby: Shown by Lightning Photography," *Strand* 12 (August 1896): 134.

6. Boleslas [Bolesław] Matuszewski, *Une nouvelle source de l'histoire* (Paris, 1898).

7. See chap. 9, pp. 192ff.

8. See chap. 4, pp. 94–96.

9. See Glasgow and Aberdeen episodes in chap. 6, pp. 186–88.

10. Wood, *Miracle of the Movies* (1947), 2. Although this is by no means a scholarly book, even by the standards of Terry Ramsaye, it is apparent that Wood knew and had interviewed Paul for his earlier *Romance of the Movies* (London: Heinemann,1937).

11. *Kine Weekly*, 8 August 1907, 195.

12. The Kinematograph Manufacturers Association was formed on 19 July 1906, with twelve companies listed as founder-members.

13. "Self-Preservation in the Trade," reproduced in Stephen Herbert, ed., *A History of Early Film* (London: Routledge, 2000), 2:397.

14. Report of KMA First Annual Meeting and Dinner, *Kine Weekly*, 7 November 1907.

15. André Gaudreault, "The Infringement of Copyright Laws and Its Effects (1900–1906)," *Framework* 29 (1985): 4.

16. Richard Brown, "The British Film Copyright Collection," *Journal of Film Preservation*, no. 54 (1997), 244n6.

17. On Edison's unscrupulous use of European films in his Vitascope screenings, see Charles Musser, *The Emergence of Cinema: The American Screen to 1907* (Berkeley: University of California Press, 1994), 118.

18. After being given at the dinner, Jago's paper was serialized in the *Kine Weekly* over four issues, presumably expanded from its first presentation.

19. *Turner v. Robinson*, 10 Ir. Ch. 121, 510. Heard at the Rolls Court in Dublin, Ireland, then part of Great Britain.

20. Additional information on this celebrated early case in photographic history comes from Heinz K. Henisch and Bridget Ann Henisch, *The Photographic Experience, 1839–1914: Images and Attitudes* (State College: Pennsylvania State University Press, 1994), 306.

21. "The Death of Chatterton," *Athenaeum*, no. 1652 (25 June 1859), 841–42.

22. "Death of Chatterton."

23. Judgment in *Turner v. Robinson*, 10 Ir. Ch. 121, 510.

24. *Edison v. Lubin*, 122 Fed. Rep. 240.

25. See Brown, "British Film Copyright Archive," 242–43.

26. See Charles Musser, *Before the Nickelodeon: Edwin S. Porter and the Edison Manufacturing Company* (Berkeley: University of California Press, 1991), 421, 552. See also a discussion of this case in L. Trotter Hardy, "Copyright and 'New-Use' Technologies," *Faculty Publications* 187 (1999), available online through the College of William and Mary Law School. I am grateful to Jon Solomon for shedding further light on the significance of this case.

27. Copyright Act, Geo.6 5(1911) c.46, part III, section 35: "'Dramatic work' includes any piece for recitation, choreographic work or entertainment in dumb show, the scenic arrangement or acting form of which is fixed in writing or otherwise, *and any cinematograph production where the arrangement or acting form or the combination of incidents represented give the work an original character*" (my italics).

28. The United States remained outside the Berne Convention until 1952, when a special Universal Copyright Convention was created to meet its requirements. This was then overtaken by America's Implementation Act in 1988, which brought the US

fully into the Berne process and led eventually to the foundation of the World Intellectual Property Organization in 1996.

29. "Standard Film Perforations," *Kine Weekly*, 5 December 1907; "The Kinematograph in Technical Education," *Kine Weekly*, 3 December 1908.

30. This seems to be the first appearance in print of an anecdote that would be transposed into the biography of William Friese-Greene, and would eventually became the basis of a scene in the *The Magic Box*. See chap. 11.

31. The British Board of Film Censors was established in 1912 to ward off potential government intervention. See http://www.bbfc.co.uk/education-resources/student-guide/bbfc-history.

32. Low and Manvell, *History*, 74.

33. Low and Manvell, *History*, 74.

34. The term "publisher" referred to distributors, who might be actual producers, although producers did not always act as their own distributors. For details of the two meetings, see Jean-Jacques Meusy, "International Meetings and Congresses of the Manufacturers Held in Paris, 1908–1909. French Viewpoints," Academia.edu/jean-jacquesMEUSY, n.d.

35. *Bioscope*, 11 March 1909.

36. *The Last Days of Pompeii* was produced by Ambrosio in 1908 (opening in Britain and the United States in 1909) at 1,200 feet. Four years later, in 1913, the same company's more spectacular remake of the same subject more ran to 6,400 feet. See Christie, "Ancient Rome in London: Classical Subjects in the Forefront of Cinema's Expansion after 1910," in *The Ancient World in Silent Cinema*, ed. P. Michelakis and M. Wyke (Cambridge: Cambridge University Press, 2013), 109–24.

37. *Kine Weekly*, 1 July 1909.

38. Paul, "Before 1910," *Proceedings of the British Kinematograph Society*, no. 38 (1936), 6–7.

39. Urban's feature-length Kinemacolor productions included *With Our King and Queen through India* (1912), running over two hours.

40. *Kine Weekly*, 20 May 1910, 94. See chap. 8, pp. 186–88.

41. A comparable incident in the life of a pioneer was Emile Reynaud destroying much of his Praxinoscope equipment, coincidentally around the same time, in 1910, after his Théâtre Optique had been rendered obsolete by the Cinématographe and its successors. Paul's closer contemporary, Méliès, would also destroy his remaining stock of films in 1923, a decade after he too ceased production.

Chapter 10

1. Edison appears to have played no direct role in the management of his two production companies, the Edison Manufacturing Co. (1894–1911) and Thomas Edison, Inc. (1911–1918), which made around 1,200 films, as compared with Paul's approximately 800 between 1896 and 1909. Likewise, Louis Lumière had little involvement in his company's production after the early years, turning his attention to color photography and later stereoscopy. Lumière production and exhibition ceased by 1902, with its last catalogue appearing in 1905, and a total of some 2,000 short films, almost all preserved.

2. M. J. G. Cattermole, *Horace Darwin's Shop: A History of the Cambridge Scientific Instrument Company 1978–1968* (Bristol: Adam Hilger, 1987), 104; G. Watchorn, *Never*

a Dull Moment: A History of the Cambridge Instrument Company 1918–1968 (Cambridge, 1990).

3. Paul is reported as "having taken additional premises at 114 and 115 Great Saffron Hill . . . fitted these with new machinery and engines, and [having] increased his staff of assistants." *Lightning*, 15 November 1894, 318. The record of his confirmation as a full member of the Institute of Electrical Engineers in December 1895 states that he employs "about thirty hands."

4. British patent 4276 of 1892, "Improvements in Electric Meters." The booklet was *Prof Ayrton and Mather's Galvanometer and Its Accessories* (1893). See C. N. Brown, "Instrument Making to Instrument Manufacturing: R. W. Paul's Unipivot Galvanometer as a Case Study," *Bulletin of the Scientific Instrument Society*, no. 71 (2001), 17.

5. Richard Brown has discovered a number of early patent applications by Paul, stretching back to the beginning of his electrical career and continuing through the Kinetoscope period. Richard Brown and Barry Anthony, *The Kinetoscope: A British History* (East Barnet: John Libbey, 2017), 53.

6. Paul, letter to Messter, 5 August 1932; Bundesarchiv-Filmarchiv, Barnes Collection.

7. Cover of Paul, *The Hundred Best Animated Photograph Films*, Season 1900–1901 catalogue.

8. There is no mention in this or any other extant catalogue of Paul offering an electrical camera, although he appears to have been using one from as early as July 1897, for filming the Jubilee.

9. *Kine Weekly*, 3 December 1908, 753–54. See also Christie, "A Scientific Instrument? Animated Photography among Other New Imaging Techniques," *The Image in Early Cinema: Form and Material*, ed. Scott Curtis, Philippe Gauthier, Tom Gunning, and Joshua Yumibe (Bloomington: Indiana University Press, 2018), 185–92.

10. Cattermole, *Horace Darwin's Shop*, xxxx.

11. Robert Williams Wood (1868–1955) would later move to Johns Hopkins University. A memoir by the Royal Society, of which he was a corresponding fellow, summed up his work and character: "Wood's active period of scientific productivity coincided with the rise of atomic physics and he made important contributions toward the increasing knowledge of the structure of the atom, chiefly through his experimental researches in physical optics. He was, however, far from one-sided and penetrated into many fields. He went wherever his insatiable curiosity led him, whether this was into different branches of physics or into all sorts of other activities such as engineering, art, crime detection, spiritualism, psychology, archaeology and many others." http://rsbm.royalsocietypublishing.org/content/2/326.

12. R. W. Wood, "The Photography of Sound-Waves and the Demonstration of the Evolutions of Reflected Wave Fronts with the Cinematograph," *Nature* 62, no. 1606 (1900): 342–49, doi:10.1038/062342a0.

13. Vernon Boys's *Soap Bubbles: Their Colours and the Forces Which Mould Them* is considered a classic of scientific popularization, first appearing in 1890 and continuously in print thereafter through a variety of publishers, starting with the Society for Promoting Christian Knowledge (SPCK) and, in recent decades, Dover.

14. The leading biologist Sir Ray Lankester, wrote in the *Bioscope* in January 1911, looking forward optimistically "to the provision, not later than next year, of a cinematograph lantern in every board school, and in every college classroom." Needless to say, he would be disappointed.

15. Brown, "Paul's Unipivot Galvanometer," 17–20.

16. Photographers would use Weston exposure meters throughout much of the twentieth century.

17. Brown, "Paul's Unipivot Galvanometer," 18.

18. Cattermole, *Horace Darwin's Shop*, 103.

19. The Patterson-Walsh device was introduced by the War Office in 1915.

20. Speech made on 10 August 1914, apparently after some days of voluntary work by staff to "overhaul the premises and prepare for a fresh start." C.S.I. Co. [Cambridge Scientific Instruments], box 33, R. W. Paul, "Address to Employees"; quoted in Cattermole, *Horace Darwin's Shop*, 102–3.

21. Clifford Copland Paterson (1879–1948) was born in Stoke Newington, son of a tanner and leather merchant. He was educated at Mill Hill School and then trained in general and electrical engineering at Finsbury Technical College and Faraday House. In 1901 he was invited to join the newly created National Physical Laboratory, where he became responsible for the Electrotechnical and Photometric Departments. He was awarded an OBE in 1916 for his work on the Paterson-Walsh aircraft height finder, and in the same year was asked by Osram Lamp Works in Hammersmith to set up their research department. He later become a founder director of the GEC research laboratories, and served as president of the IES (1928–1929) and the IEE (1930–1931).

22. From Clifford Paterson, *A Scientist's War: The War Diary of Sir Clifford Paterson 1939–45*, ed. Clayton and Algar (London: Peter Peregrinus, 1991), quoted in Cattermole, *Horace Darwin's Shop*, 103.

23. A "slop-pail" was a form of chamber pot. A. B. Wood, "From Board of Invention and Research to Royal Naval Scientific Servoce," *Journal of the Royal Naval Scientific Service* 20, no. 4 (1965): 31.

24. Bragg was awarded the CBE in 1917 for his war work, and he was knighted in 1920. Having been a fellow since 1907, he was elected president of the Royal Society in 1935.

25. Cattermole, *Horace Darwin's Shop*, 104.

26. "Darwin, Horace (DRWN868H)," *Cambridge Alumni Database*, University of Cambridge. Also Cattermole, *Horace Darwin's Shop*.

27. Watchorn, *Never a Dull Moment*, 9.

28. Watchorn, *Never a Dull Moment*, 16. The vague chronology of this memoir makes it difficult to date its observations accurately.

29. While researching Paul, I was contacted by Don Unwin, a retired employee of the Cambridge Instrument Company, who published a short history of it and had bought a milling machine that once belonged to Paul, which he was kind enough to show me. Donald J. Unwin, *"The Scientific": The Story of the Cambridge Instrument Company* (Cambridge Industrial Society, 2001).

30. William Henry Bragg, RI Admin Correspondence 1933–1939, Royal Institution of Great Britain. Bragg's neighbor may well have been in the Surrey village of Chiddingfold, where he had a country house, which would explain the need for a solution not requiring electricity.

31. The fullest account of the Pulsator and its uses and modifications is "The Pulsator: How a Portable Artificial Respirator Saved the Lives of Children," https://fromthehandsofquacks.com/2016/05/04/the-pulsator-how-a-portable-artificial-respirator-saved-the-lives-of-children/.

32. R. W. Paul, "The Bragg-Paul Pulsator," *Proceedings of the Royal Society of Medicine* 28, no. 4 (February 1935): 436–38.

33. "The Pulsator."

34. "The Pulsator."

35. "Cineradiographic Films Illustrating Normal and Artificial Respiration with the Bragg-Paul Pulsator," *Journal of Physiology*, 1 December 1939.

Chapter 11

1. Although Friese-Green filed the patent, his apparatus was judged to have not yet achieved what he promised, or hoped. However, despite much controversy over his achievements, revisionist scholarship now claims that he succeeded better than was believed after a sharply critical study by Brian Coe, "William Friese-Greene and the Origins of Cinematography," *Screen* 10, nos. 2–4 (1969). See Peter Domankiewicz's blog, https://friesegreene.com/.

2. See Luke McKernan, *Charles Urban: Pioneering the Non-Fiction Film in Britain and America, 1897–1925* (Exeter: University of Exeter Press, 2013), 118–24. McKernan evokes the "baleful presence" of Friese-Greene and his unsupported claims, which led to a high court judgment in 1914 ending Urban's fragile "monopoly on colour."

3. My thanks are due to Peter Domankiewicz for making available the earliest example of Day insisting on Friese-Green having "invented" film, in an interview quoted on 7 May 1921. Describing Friese-Green as "one of England's greatest unrecognised inventors," and his life history as "full of tragic romance," Day laid the foundations for what would later be regarded as the true history of Friese-Greene anticipating Edison with his 1890 patent. See "Inventor of the Cinema," *Shields Daily News*, 7 May 1921, and the section "Paul's Afterlife" in this chapter.

4. See the British Cinema and Television Veterans homepage, https://www.britishcinemaandtelevisionveterans.org.uk/about-us/our-history/.

5. This criterion was updated a number of times in subsequent years, with television added in 1978.

6. Correspondence between Paul and Will Day confirms that they remained in contact, with a common interest in the welfare of some veterans. I am grateful to Laurent Mannoni, curator at the Cinémathèque Française, for access to the Will Day collection.

7. In a vigorous contribution to the discussion of a paper on "scientific management" by J. M. Scott Maxwell, which advocated adoption of F. W. Taylor's method of measuring and rewarding work by splitting a job into its elements—practices that came to be known as "Taylorism"—Paul attacked the approach as "unscientific . . . because it confuses work and effort." His speech advocated viewing labor relations "more from the workman's point of view," and paying wages "on a juster basis that will enable workers to pay for what they need." "Discussion on 'Scientific management: a solution of the capital and labour problem' before the Institution, 11 December, 1919," *Journal of the Institution of Electrical Engineers* 58, no. 291 (May 1920): 342–76, https://digital-library.theiet.org/content/journals/10.1049/jiee-1.1920.0024 (11.4.19).

8. For Physical Society and Institute of Physics history, see http://www.iop.org/about/history/page_38385.html.

9. The Royal Institution Christmas Lectures were broadcast on BBC television

from 1966 to 1999, and subsequently had a checkered history on different channels, before returning to BBC television in 2010.

10. W. H. Eccles, "Robert W. Paul: Pioneer Instrument Maker and Cinematographer," *Electronic Engineering*, August 1943.

11. Faraday Exhibition souvenir catalogue (London: Royal Albert Hall, 1931), 7.

12. The introduction is signed "W. H. B.," indicating that it was written by Sir William Bragg, one of the prime movers.

13. Faraday Exhibition catalogue, 17.

14. Robert W. Paul, "Some Electrical Instruments at the Faraday Centenary Exhibition, 1931," *Journal of Scientific Instruments* 8:1

15. Paul, "Some Electrical Instruments," 2.

16. The Paul Instrument Fund, now administered by the Royal Society, provides up to £75,000 to develop "genuinely new ideas, techniques or highly novel applications of existing ideas and techniques. The instrument should, as far as possible, be a stand-alone device and might be an outcome of a previous extensive research program. The scheme aims to support innovative development, rather than pure research."

17. The bibliography for his article ranges from 1784 to 1930.

18. Frederick A. Talbot, *Moving Pictures: How They Are Made and Worked* (London: Heinemann, 1912). Talbot's book was republished in 1923, largely unchanged, except—as Richard Brown has observed—a crucial phrase was omitted, concerning Paul's right to manufacture Kinetoscopes in 1894. Brown, "A New Look at Old History—the Kinetoscope: Fraud and Market Development in Britain in 1895," *Early Popular Visual Culture* 10, no. 4 (November 2012): 412.

19. Terry Ramsaye, *A Million and One Nights: A History of the Motion Picture through 1926* (New York: Simon & Schuster, 1926), 149–50.

20. Much of this "complication" swirled around Edison, who provided the frontispiece and an endorsement for Ramsaye's book as "the first endeavour to set down the whole and true story of the motion picture." Day's insistent advocacy of the 1890 Friese-Greene patent would continue to trouble Edison until the 1930s.

21. Ramsaye, *Million and One Nights*, 159.

22. H. G. Wells, *The King Who Was a King: The Book of the Film* (London: Ernest Benn, 1929), 10.

23. Oskar Messter (1866–1943), letter to "Mr Robert W. Paul, Manufacturer of Kinematographic Instruments, London," 14 July 1932; Bundesarchiv-Filmarchiv, Barnes Collection.

24. Paul, letter to Messter, 5 August 1932.

25. Paul, letter to Messter, 30 January 1935. The "friend" mentioned would presumably have been Will Day.

26. Oskar Messter, *Mein Weg mit dem Film* (Berlin: Max Hesse, 1936), 150. From this illustration, John Barnes diagnosed Messter's machine as based on an "imperfect imitation" of a reconstruction of Paul's second Theatrograph (Mark 1). Barnes, *The Beginnings of the Cinema in England* (1976; University of Exeter Press, 1998),1:206.

27. Stephen Barber, "The Skladanowsky Brothers: The Devil Knows," *Senses of Cinema*, no. 56 (October 2010), http://sensesofcinema.com/2010/feature-articles/the-skladanowsky-brothers-the-devil-knows/.

28. Rachael Low, *Film Making in 1930s Britain* (Allen & Unwin, 1985), 22.

29. The organization was originally the "London Branch" of the Society of Motion Picture Engineers (now the Society of Motion Picture and Television Engineers). It

became the British Kinematograph Society in 1931, and began accrediting courses at the London Polytechnic in the following year. In the 1970s, "Sound and Television" were added to the Society's name, and since 2016 it has been known as the International Moving Image Society.

30. The use of slides and extracts indicates that Paul must have kept at least some material from his film career.

31. Paul, "Before 1910," *Proceedings of the British Kinematograph Society*, no. 38 (1936), 5.

32. Paul, "Before 1900," 6. On *The Magic Sword*, see chap. 8, pp. 174–76.

33. In his presentation, Hepworth showed a group portrait of the Paris conference delegates (including "the Chief Conspirator R. W. Paul . . . Nice looking fellow—in those days") and described how the exhibitors "took fright" at the Paris scheme to fix rental prices, thus paving the way for "the American invasion." "Before 1910," 11.

34. "Before 1910," 7.

35. The BBC's Broadcasting House had opened in 1932, and its experimental Television Service was launched on 2 November 1936, operating from a wing of Alexandra Palace in North London, less than a mile from the site of Paul's original Muswell Hill studio, still serving as his factory at this time.

36. Known as the Polytechnic Theatre at this period, the hall would undergo various name changes and refittings, mostly incorporating the name "Cameo," between the 1940s and 1980s, before its most recent incarnation as the Regent Street Cinema in 2016.

37. The others mentioned by name were Matt Raymond, Albany Ward, G. D. Adams, and Will Barker.

38. Westminster University website, https://www.westminster.ac.uk/news-and -events/news/2012/the-birth-of-british-cinema-lumières-cinematograph.

39. Will Day's collection of early film apparatus was on display at the South Kensington Museum (forerunner of the Science Museum) for much of the 1920s, before it was offered for sale and eventually acquired by the Cinématheque Française in 1959, together with his papers. See details in Stephen Bottomore, "Will Day: The Story of a Discovery," *Film Studies*, no. 1 (Spring 1999), 81–90.

40. Wells's "dream of the future," *The Shape of Things to Come*, appeared in 1933, predicting the outbreak of a devastating war in 1940. His partial adaptation of the novel for Korda served as the basis for an elaborate display of special effects and modernist design, involving, among others, László Moholy-Nagy and Fernand Leger, with the American production designer William Cameron Menzies credited as director.

41. Leslie Wood, *The Miracle of the Movies* (London: Burke Publishing, 1947), 106.

42. Watchorn, *Never a Dull Moment*, 16. From the comments that follow this, on Chaplin and Pickford, it seems likely that Watchorn was recalling, or confusing, a conversation from the 1920s.

43. R. W. Paul, "Kinematographic Experiences," *Journal of the Society of Motion Picture Engineers* 27, no. 5 (1936): 498.

44. Gutsche's doctoral thesis, "The History and Social Significance of Motion Pictures in South Africa 1895–1940," was completed in 1946, and later appeared as a book of the same title (Cape Town, H. Timmins, 1972).

45. Paul, letter dated 20 September 1937; Bundesarchiv-Filmarchiv, Barnes Collection.

46. Cattermole, *Horace Darwin's Shop*, 112.

47. Cattermole, *Horace Darwin's Shop*, 142.

48. Paul, letter to Whipple, 25 December 1941; quoted in Cattermole, *Horace Darwin's Shop*, 143.

49. https://royalsociety.org/grants-schemes-awards/grants/paul-instrument/.

50. Rachael Low had served in the Treasury during the war, and was the daughter of David Low, a famous cartoonist and creator of the Colonel Blimp character, best known today through Michael Powell and Emeric Pressburger's controversial film *The Life and Death of Colonel Blimp* (1943).

51. Terry Ramsaye, review of *Friese-Greene: Close-Up of an Inventor*, in *Journal of the Society of Motion Picture Engineers* 52, no. 4 (April 1949).

52. A series of articles by Brian Coe, curator of the Kodak Museum in Britain, in the 1950s and 1960s challenged much that had been claimed by Forth, and before her, Will Day. See Coe, "The Truth about Friese Greene," *British Journal of Photography* (1955), and later articles in *Screen*.

53. Wood, *Romance of the Movies*, 27.

54. Low and Manvell, *History*, 23.

55. A later reprint of Low's history, by Routledge in 1997, failed to correct the "1900" misprint.

56. Low and Manvell, *History*, 24.

57. The second volume covered 1906–1914, and the third 1914–1918, all three appearing in 1948. Hepworth, who chaired the BFI Research Committee, did not die until 1953.

58. Lindgren's account of cinema aesthetics was strongly influenced by prevailing concepts, including the primacy of "montage," as developed by Griffith and Eisenstein and carried into the shot–reverse shot structures of the sound era. This conception of "film language" would later be characterized by David Bordwell as "the basic story," or the "narrative that traces the emergence of film as a distinct art." Bordwell, *On the History of Film Style* (Cambridge, MA: Harvard University Press, 1987), 12–13 ff. Lindgren's book *The Art of the Film: An Introduction to Film Appreciation* first appeared in 1948 (London: Allen & Unwin) and has gone through several later editions.

59. Low and Manvell, *History*, 52.

60. Low and Manvell, *History*, 56. It is hard to believe that Low would have seen any of the *Army Life* series, only one fragment of which is known to survive.

61. Low and Manvell, *History*, 86.

62. Low and Manvell, *History*, 86.

63. It seems likely that few of Paul's films were available by the late 1940s, although eight of the twenty-three listed as "preserved by the National Film library" are his. *The ? Motorist* is listed, and three stills appear in the book, although there is no discussion of the film, and Paul's major multiscene drama, *Scrooge* (1901), is not mentioned.

64. Georges Sadoul, *Histoire générale du cinéma*, vol. 2, *Les pionniers du cinéma (de Méliès à Pathé 1897–1909)* (Paris: Denoël, 1948). Sadoul's history was the first multivolume comparative study, based on actual inspection of documents and surviving films, since the earlier synoptic histories that followed Ramsaye, such as Bardèche and Brasillach's, *History of the Film* (Paris: Denoel et Steele, 1935), which appeared in English translation in 1938. Sadoul's original edition was revised for republication in 1973 by Bernard Eisenschitz.

65. Sadoul, *Histoire*, 2:129. For no obvious reason, Sadoul always referred to Paul as "William."

66. Sadoul, *Histoire*, 2:130–31.

67. Sadoul, *Histoire*, 2:137–38

68. Sadoul, *Histoire*, 2:171.

69. Sadoul, *Histoire*, 2:172. Sadoul acknowledged his debt to Low and Manvell, who had shown him their manuscript for *History of the British Film* before its publication. This presumably lies behind his adoption of the same harsh verdict on Paul's "vulgarity," although it hardly explains it.

70. Could this in fact be a confused reference to the 1910 film by James Williamson, *The History of a Butterfly: A Romance of Insect Life*, which was his last?

71. While the importance of the 1978 congress is not denied, Philippe Gauthier observes that it was one of a number of events that created "the new film history"; https://www.academia.edu/1795860.

72. This would later become, in revised form, the first in Barnes's five-volume series, reaching 1900. References here are to the 1998 University of Exeter Press edition.

73. Barnes, *Cinema in England*, 1:xiv.

74. Richard Brown, "England's First Film Shows," *British Journal of Photography*, 31 March 1978, 273. Brown's earlier article had appeared in the *BJP*, 24 June 1977, and stimulated a vigorous correspondence, involving a dissenting Dutch historian Tjitte de Vries and the English photographic historian Baynham Honri (12 August 1977).

75. See, for example, Simon Popple, "'But the Khaki-Coloured Camera Is the Latest Thing': The Boer War Cinema and Visual Culture in Britain," in *Young and Innocent? The Cinema in Britain 1896–1930*, ed. Andrew Higson (Exeter: University of Exeter Press, 2002), 13–27, and Stephen Bottomore's contribution to Andrew Shail, ed., *Reading the Cinematograph: The Cinema in British Short Fiction, 1896–1912* (Exeter: Exeter University Press, 2011).

76. Michael Chanan, *The Dream That Kicks: The Prehistory and Early Years of Cinema in Britain* (London: Routledge, 1980).

77. Barry Salt, *Film Style and Technology: History and Analysis* (London Starword, 1983). Conspicuously missing from the films Salt discusses are a number of innovative Paul titles from 1899–1901, such as *The Countryman and the Cinematograph*, *Scrooge*, and *The Magic Sword*. Since these are also omitted from Low, it may be that they were not available to view until more recently; although in 1991 Musser cited *The Countryman* as the precursor of Edison's *Uncle Josh at the Moving Picture Show*.

78. Noel Burch, *Life to Those Shadows* (London: British Film Institute, 1990).

79. Burch, *Life to Those Shadows*, 92.

80. Emmanuelle Toulet, *Cinématographe, invention du siècle* (Paris: Decouvertes Gallimard, 1988), 95.

81. Simon Popple and Joe Kember, *Early Cinema: From Factory Gate to Dream Factory* (London: Wallflower, 2004), 28.

82. Deac Rossell, *Living Pictures: The Origins of the Movies* (Albany: State University of New York Press, 1998), 140.

83. The most important of these were probably J. H. Martin, G. H. Cricks, Walter Booth, Frank Mottershaw, and Jack Smith, although others await identification.

84. See, for example, two references to Paul films in Sarah Street, *British National*

Cinema (Abingdon: Routledge, 1997), 34. Both *An Extraordinary Cab Accident* and *The ? Motorist* are attributed to Booth by the convention of giving the director after the title. In IMDb, some forty Paul films are now attributed to Booth as director.

85. Barnes, *Cinema in England*, 5:15.

86. Program, Giornate del cinema muto, Pordenone, 2016. Barnes's article was published posthumously by the Magic Lantern Society (newsletter 92).

Epilogue

1. Some 830 titles were recovered in Blackburn and preserved by the BFI National Archive, in partnership with the National Fairground and Circus Archive, University of Sheffield. See https://web.archive.org/web/20060313205229/http://www.bfi.org.uk/features/mk/.

2. Simon Brown, "Flicker Alley: Cecil Court and the Emergence of the British Film Industry," *Film Studies* 10 (2007): 21–33.

3. For the early period, covered in magisterial detail up to 1860, see Richard Altick, *The Shows of London* (Cambridge, MA: Harvard University Press, 1978). For a sketch of later developments, see Christie, "Screening the City: A Sketch for the Long History of London Screen Entertainment," http://ianchristie.org/screening the city.html.

4. See Stephen Herbert, *Industry, Liberty, and a Vision: Wordsworth Donisthorpe's Kinesigraph*, 2nd ed. (Hastings: Projection Box, 2018).

5. Some twenty-eight hours of Mitchell & Kenyon films are now restored, while Paul's surviving work runs less than three hours.

6. Low and Manvell, *History*, 1:13.

7. See, for instance, Stefan Collini, "Snow v Leavis: The Two Cultures Bust-Up 50 Years On," *Guardian*, 16 August 2013, https://www.theguardian.com/books/2013/aug/16/leavis-snow-two-cultures-bust.

8. The legacy of F. R. Leavis and his 1930s journal *Scrutiny* has long been invoked in discussion of attitudes to popular culture in British academia. For an overview, see F. R. Leavis and Denys Thompson, *Culture and Environment: The Training of Critical Awareness* (London: Chatto and Windus, 1959).

9. *Times Literary Supplement* exchange over *Silence*, May 2017. Also Christie, "Pictures and Prejudice: The Origins of British Resistance to Film," *Times Literary Supplement*, 17 November 1995.

10. Sims's and Kipling's stories are discussed in chapter 6, Bely's essay in chapter 8. In Gorky's story, written after seeing the Lumières' *Baby's Breakfast* in Nizhny Novgorod, a chorus girl tries to commit suicide in despair at the gulf between her own situation and that of the idyllic family shown in the film; see Tsivian, *Early Cinema in Russia* (Abingdon: Routledge, 1991), 36. In Normand's, published in *L'illustration*, no. 2974 (24 February 1900), an Irish girl working in Paris sees a film apparently showing her lover, a British soldier, being killed in battle in South Africa. When she bursts into tears, a neighbor in the audience chides her: "Don't you realise the soldiers you saw were mere actors? These scenes were not cinematographed in Africa. It was a bad pantomime played in Paris itself, at the Buttes-Chaumont. I can show you the place. Do you really think that photographers would take pictures under hails of bullets and cannon balls?" Apparently an English translation of this story appeared in *Soldiers of the Queen*, no. 80 (March 1895), and a German version in *Frankfurter Zeitung*, 8 July

1900, 1–3; see Stephen Bottomore, "Filming, Faking and Propaganda: The Origins of the War Film, 1897–1902" (doctoral thesis, Utrecht University, 2007), chap. 10, p. 40.

11. See chap. 8, p. 169.

12. Boxing matches were an early attraction in moving pictures. Dickson filmed the Leonard-Cushing fight in June 1894, and Corbett and Courtney were filmed over six one-minute rounds for the Kinetoscope in November 1894. In 1899 American Mutoscope and Biograph filmed the entire Jefferson-Sharkey fight, lasting over two hours. "Passion plays" were another specialized genre that depended on scale for their impact. Lubin distributed the first of his *Passion Play* films in 1898, followed by an hour-long version in 1903.

13. Paul's film was clearly copied in Edison's *Uncle Josh at the Moving Picture Show* (1902). Later films set in cinemas included *Those Awful Hats* (Biograph, 1911).

14. Not only greater accuracy but complex lines of authority and trust linked to such powerful figures as Paul's teacher, Ayrton. See Graeme J. N. Gooday, *The Morals of Measurement. Accuracy, Irony and Trust in Late Victorian Electrical Practice* (Cambridge: Cambridge University Press, 2004), 20–21.

15. In 1914 London "had fifty-nine independent distributors of electricity, supplying current in seventy-two different varieties," with voltages "480 down to 110, and frequencies from 100 down to 25"; Gavin Weightman, *Children of Light* (London: Atlantic Books, 2011), 143.

16. See pp. 209–10 for references to Paul's speech to his workers in 1914.

17. Herkomer, quoted in "Trade Topics," *Bioscope*, 30 October 1913, 341.

18. None of Herkomer's five films are known to survive. See Lynda Nead, "Paintings, Film and Fast Cars: A Case Study of Hubert von Herkomer," *Art History* 25, no. 2 (2003): 240–55.

19. Lynda Nead, *The Haunted Gallery: Painting, Photography, Film c. 1900* (New Haven, CT: Yale University Press, 2007), 200.

20. Nead unfortunately does not refer to Paul.

21. Herkomer, *Bioscope*, 1913

22. Tom Gunning, "The Cinema of Attraction: Early Cinema, Its Spectator and the Avant-Garde," *Wide Angle* 8, no. 3 (1986): 63–70; André Gaudreault, *Film and Attraction: From Kinematography to Cinema* (Urbana: University of Illinois Press, 2011). See also Wanda Strauven, ed., *The Cinema of Attractions Reloaded* (Amsterdam: University of Amsterdam Press, 2006).

23. In *Kine Weekly*, 1907. See chap. 6, p. 141.

24. Tsivian, *Early Cinema in Russia*, 34–35. Although this comes from a source dated 1911, Bely's article appears to have been written in 1907.

25. Tsivian, *Early Cinema in Russia*, 35.

26. H. G. Wells, "The Man of Science as Aristocrat," *Nature* 147 (19 April 1941): 467.

27. Paul to the editor, *Nature*, 17 May 1941.

28. Bernard Bergonzi, *The Early H. G. Wells* (Toronto: University of Toronto Press, 1961). See also David Lodge, "Assessing H. G. Wells," *Encounter* 27, no. 1 (January 1967): 55.

29. On Kahn, see Teresa Castro, "Les archives de la planète d'Albert Kahn," in *Lieux de savoir: Espaces et communautés*, ed. Christian Jacob (Paris: Albin Michel, 2007); also "Les Archives de la Planete: A Cinematic Atlas," *Jump Cut*, https://www.ejumpcut.org/archive/jc48.2006/KahnAtlas/.

30. Not enough attention has been given to what I am calling the "territorial" dimension of early film's appeal: how it enabled viewers who had often no opportunity to travel beyond their immediate community to "see the world." Modern historians of early film are always in danger of ignoring, or taking for granted, this primary appeal.

31. On linking Albert Einstein's and Eugene Minkowski's theories, see Stephen Kern, *The Culture of Time and Space, 1880–1918* (Cambridge, MA: Harvard University Press, 2003), 207.

32. Mikhail Bakhtin, "Forms of Time and of the Chronotope in the Novel," in *The Dialogic Imagination* (Austin: University of Texas Press, 1981), 84–85.

33. *A Collier's Life* was found by Camille Blot-Wellens in the Swedish Film Institute archive and restored as a joint project with the BFI National Archive. The other Swedish discovery was *The Fatal Hand*.

34. See the most recent revised entry on Demetrius Anastis Georgiades and George John Trajidis, by Theodoros Natsinas (2018), in *Who's Who in Victorian Cinema*, http://www.victorian-cinema.net/georgiades.php (accessed 8 July 2019).

35. I owe this very recent discovery to Rachael Macdonald, of the Hornsey Historical Society, and confirmation to the Royal Society archivist Jonathan Bushell.

Index

Page numbers in italics refer to figures.